顶级餐厅设计集成 · 本书编委会编

New Top Restaurants

悦食
Delightful Restaurant III

中国林业出版社
China Forestry Publishing House

图书在版编目（CIP）数据

悦食：顶级餐厅设计集成.3：汉英对照 /《悦食: 顶级餐厅设计集成》编委会编写. -- 北京：中国林业出版社, 2013.3
ISBN 978-7-5038-6981-5

Ⅰ.①悦… Ⅱ.①悦… Ⅲ.①餐馆－室内装饰设计－图集 Ⅳ.①TU247.3-64

中国版本图书馆CIP数据核字(2013)第043218号

悦食Ⅲ——顶级餐厅设计集成

◎ 编委会成员名单

主　　编：刘　晶
副 主 编：刘增强
编写成员：孔新民　贾　刚　高囡囡　王　超　刘　杰　孙　宇　李一茹
　　　　　姜　琳　赵天一　李成伟　王琳琳　王为伟　李金斤　王明明
　　　　　石　芳　王　博　徐　健　齐　碧　阮秋艳　王　野　刘　洋
　　　　　陈圆圆　陈科深　吴宜泽　沈洪丹　韩秀夫　牟婷婷　朱　博
　　　　　宁　爽　刘　帅　宋晓威　陈书争　高晓欣　包玲利　郭海娇
　　　　　张　雷　张文媛　陆　露　何海珍　刘　婕　夏　雪　王　娟
　　　　　黄　丽　程艳平　高丽媚　汪三红　肖　聪　张雨来　韩培培
采　　编：柳素荣

中国林业出版社 · 建筑与家居出版中心
出版咨询：(010) 8322 5283

出版：中国林业出版社　（100009 北京西城区德内大街刘海胡同7号）
网址：www.cfph.com.cn
E-mail：cfphz@public.bta.net.cn
电话：(010) 8322 3051
发行：新华书店
印刷：北京利丰雅高长城印刷有限公司
版次：2013年4月第1版
印次：2013年4月第1次
开本：230mm×300mm　1/16
印张：20
字数：180千字
定价：320.00元（USD 60.00）

Preface·序

设计指挥家

　　写这篇序言的此刻,刚从美国洛杉矶返回北京。卸下繁忙,坐在工作室一面敲着键盘,一面轻松小酌Penfolds Cellar Reserve 2007年的pinot noir。随着越来越密集的工作行程及频繁的差旅飞行,每年超过30个遍及全球的品牌餐饮项目设计,现在我更明白"慢活"对身心平衡的重要,只要工作一有空档,一杯好酒搭配阅读一本好书,代表着一种从容不迫、细腻的生活态度。

　　上月在新加坡发现一间外观抢眼叫"Browhaus"的店,一开始因为发现跟德国"Bauhaus"谐音,走进看竟是一家"修眉"的连锁品牌,专业的技术依据每个人脸型及轮廓,打造最符合个人形象的眉型。就连英国Wallpaper杂志也推荐为"国际旅游人士必到的修眉hotspot"。听说目前上海也有分店,也许下次我能发表变身型男的感受。

　　餐饮空间是人与人聚集交流的场所,所以作为餐饮空间设计者的我,时常会想着什么人来这餐厅,什么是来此的人所需要的,这个过程其实令我感到最有意思,就像是北方人在家包饺子,重点在于包饺子的过程人与人之间的交流,交流的好结果就是水到渠成,交流的不好水饺可能就索然无味了。

　　餐饮空间设计,是一种需要生活经验累积和内在沉淀后的表现。设计工作本身也是有趣的学习过程,通过对生活体验的折射,产生源源不绝的创意灵感并激发出对生命更丰富的热情。像是艺术家或音乐家,基本同样的画笔或是乐器,透过各自精采性格与洋溢才华的表现,总是能撩拨起你我情绪中的那根敏感的神经,而整合了国内外优秀创意的"悦食3",理所当然的成为整合一切美好设计协奏曲的指挥家。

2013年1月10日

THE SOUTH BEAUTY RESTAURANT, TAIPEI / 8
CHOWHAUS AT HUASHAN RD. / 14
CUISINE CUISINE BEIJING / 22
MAZZO AMSTERDAM / 30
UNICO+COLAGRECO / 38
SABATINI-SH / 54
SESAME JAPANESE RESTAURANT / 64
BEIJING KINGSJOY / 70
SABURI ATT 4 FUN / 78
YAKINIKU MASTER JAPANESE BARBECUE RESTAURANT / 86
DESIGN DESCRIPTION OF THE INKED ORCHID-PAVILION / 92
WALNEW CLUB CHANGGUANG WETLAND PARK, WUXI / 98
HBO MOVIE THEME RESTAURANT / 108
JIULI RIVER RESTAURANT / 114
XISI TEPPANYAKI / 124
STARRY NIGHT DINING / 130
KARUIZAWA ON GONGYI ROAD / 136
SIMPLICITY RESTAURANT / 146
STARRY NIGHT DINING / 152
YANGZHOU FULIN HU GARDEN / 160
ASSAGGIO TRATTORIA ITALIANA / 166

Contents·目录

HOTEL CASTELL D'EMPORDA /172

BEIJING FINANCIAL STREET CONTINENTAL HOTEL XIN RONG JI RESTAURANT /178

SPRING FEAST /186

CHARME RESTAURANT A DREAM OF DRAGON AT HONGKOU, SHANGHAI /194

WONG'S CHAFING DISH RESTAURANT AT YIZHUANG /200

BEIJING LONGTAN LAKE PARAMOUNT CHAMBER /210

SPICE SPIRIT RESTAURANT /218

MEIZHOU DONGPO RESTAURANT, YIZHUANG STORE /224

MANGO THAI IN NINGBO /232

CLUB HOUSE OF HONOR, TONGREN, GUIZHOU /238

DESIGN DESCRIPTION OF THE FAVORITE PAVILION /250

BIANYI WORKSHOP /256

XIANG HE BAI NIAN /264

LITTLE COOK SEAFOOD RESTAURANT /274

TONGQING BUILDING LUCHOU FU /280

YONGXIANG FASHIONABLE RESTAURANT /288

CO-EXIST & HARMONIOUS /294

IMPRESSION OF WANG JIANG NAN /304

IZAKAYA SINGER /310

SANYA QIXIAN RIDGE WESTERN RESTAURANT /316

台北俏江南餐厅 / 8

华山路 Chowhaus 餐厅 / 14

北京国锦轩 / 22

Mazzo 餐厅 / 30

唯壹餐厅 / 38

Sabatini-SH 餐厅 / 54

Sesame 日式餐厅 / 64

北京京兆尹 / 70

纱舞缤 ATT 4 FUN 店 / 78

烧肉达人日式烧肉店 / 86

水墨兰亭 / 92

无锡长广溪湿地公园蜗牛坊 / 98

HBO 电影主题餐馆 / 108

九里河餐厅 / 114

西四铁板烧 / 124

正村寿司 HK-LokFu 店 / 130

轻井泽公益店 / 136

朴素餐厅 / 146

星光捌号 / 152

扬州富临壶园府邸 / 160

Assaggio Trattoria 意大利餐厅 / 166

Contents · 目录

castell d'emporda 餐厅 / **172**

北京金融街洲际酒店新荣记餐厅 / **178**

春江宴 / **186**

港丽餐厅上海虹口龙之梦店 / **194**

王家渡火锅亦庄店 / **200**

龙潭湖九五书院 / **210**

麻辣诱惑上海虹口龙之梦店 / **218**

眉州东坡酒楼亦庄店 / **224**

宁波美泰泰国餐厅 / **232**

贵州铜仁上座会馆 / **238**

所好轩(沌口店) / **250**

便宜坊 / **256**

祥和百年 / **264**

小厨师海鲜餐厅 / **274**

同庆楼庐州府 / **280**

涌香格调时尚餐厅 / **288**

客家本色大里店 / **294**

印象望江南 / **304**

圣家居酒屋 / **310**

三亚七仙岭西餐厅 / **316**

Johannes Torpe
Johannes Torpe Studios 设计总监
丹麦著名设计师

The South Beauty Restaurant, Taipei
台北俏江南餐厅

设计单位：Johannes Torpe Studios
项目地点：台北
竣工时间：2012

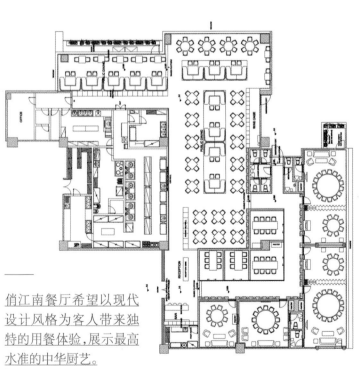

俏江南餐厅希望以现代设计风格为客人带来独特的用餐体验，展示最高水准的中华厨艺。

　　餐厅的家具是由我们和Johanners Torpe工作室特别设计的。为了能给客人带来最美妙奢华的用餐环境，我们在设计的各个方面都很用心。

　　厨房对面是一个LED显示屏，里面的画面是一个流动的瀑布，透过屏幕你可以看到大厨们忙碌的灶火，整个画面构成了中国传统风水学里重要的水与火两大元素。俏江南的天花板是一幅优美的大型水彩画。它是由许多画家通过数字手段手工绘制而成，总面积442平方米。置身画下，令人叹为观止。除了这些出类拔萃的设计，我们还为餐厅特别打造了一顶"高帽子"，犹如画龙点睛。

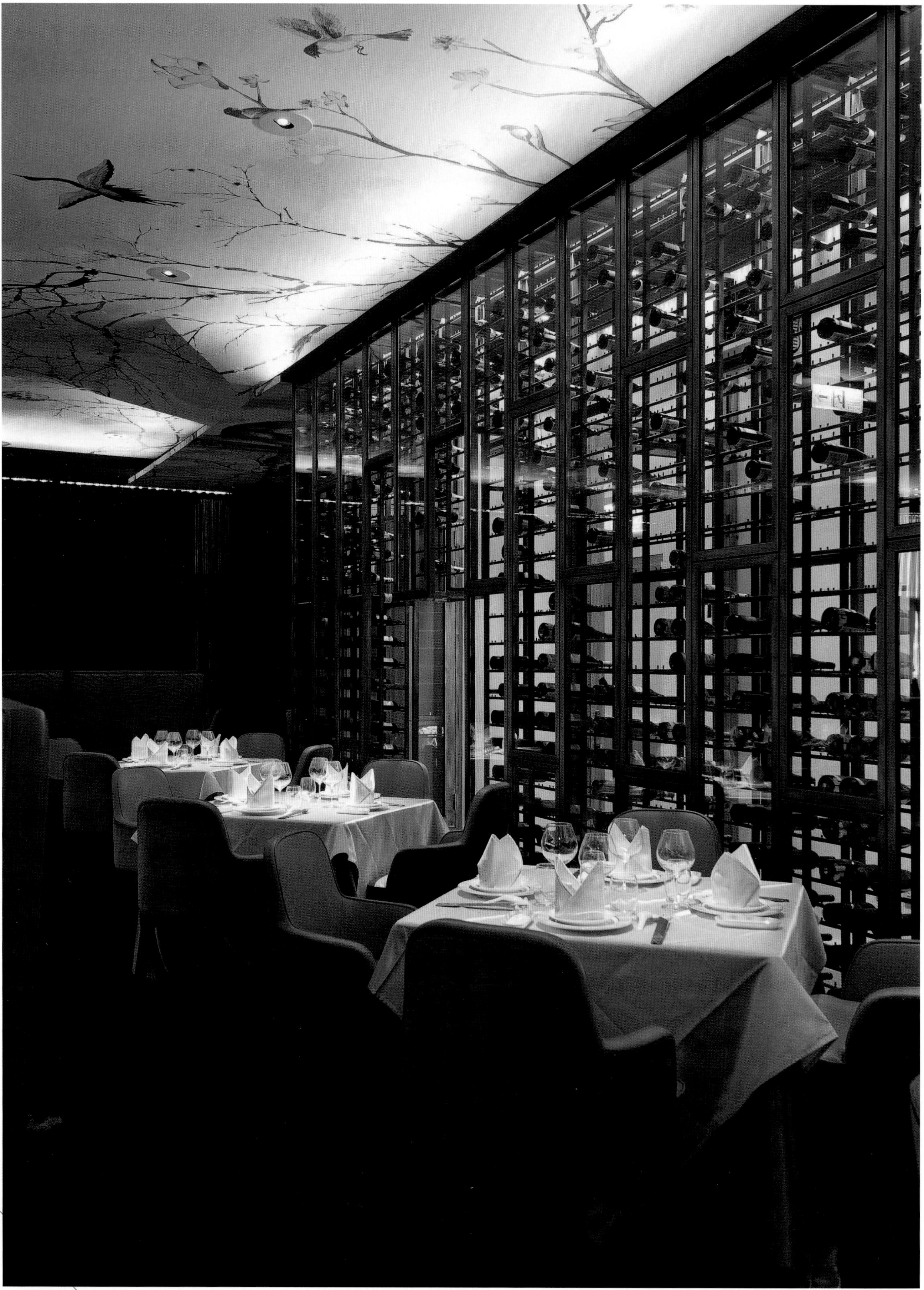

The idea behind the South Beauty restaurant was to create a truly unique dining experience, presenting the highest quality Chinese Cuisine to the world presented in a contemporary design style. We, at Johannes Torpe Studio, designed a special line of furniture to fit in to the restaurant. The vision for South Beauty Taipei was to create a restaurant, which will give the guests the most fantastic and grand dining experience. By designing all aspects of the restaurant it has been possible to do that.

Standing in front of the kitchen is a large LED screen on the entire wall, displaying a video of a continuous waterfall. When looking through this screen into the kitchen you can see the raw flames created by the chefs in action, resulting in the combination of fire and water; two key elements of the traditional Chinese Feng Shui ideology. The ceiling of South Beauty is also very special, bringing hand crafted watercolour paintings from various artists together with digital means to create a 442 square metre artwork that sits elegantly above your head, and contributes to a beautiful and breath taking experience. Together with the ceiling and the rest of the specially designed interior, the cool "high hat" lamps designed exclusively for South Beauty will complete the dining experience.

颜呈勋

教育背景

1999-2003年 美国 哈佛大学设计学院毕业(建筑系学硕士)

1994-1998年 美国 杜克大学(经济&艺术史学士学位,优等成绩;优秀生榜学员)

实践经历

2005年-至今 上海穆哈地设计咨询(MRT design),总监建筑师

2008-2010年 上海 MRKT产品,创意总监

2002-2003年 波士顿 Kennedy and Violich Architecture 事务所,建筑师

2000-2001年 纽约 Tsao and McKown Architects 事务所,建筑师

1998-1999年 波士顿 Iconomy.com,设计师

学校职务

2011年 香港大学(上海分校)讲师

2003年 上海同济大学 都市规划部 访问学者

Chowhaus at Huashan Rd.
华山路Chowhaus餐厅

设计单位:穆哈地设计咨询(上海)有限公司\MRT DESIGN

设 计 师:Bill Yen\颜呈勋

项目地点:上海

建筑面积:600 m²

竣工时间:2011

摄 影 师:MOSEMAN ELEANOR ELIZABETH

Chowhaus周边有绿树围绕,门面并不大,木质的外观显得自然低调。

餐厅的开放空间被分成4个区,右边与中间是适合午餐的座位,左侧沙发、小圆桌适合小酌,一般情况下这就是餐厅开放区的全部,可Chowhaus别有洞天。最左往里走,是玻璃房,除却放了不少植物外,中间有个装了壁炉的书架,从各地搜回的老皮箱、旅游手册成为摆设,原色木几周围放着米色、灰色的沙发,自然光线充足,看起来更像个独立的咖啡室。

三个从屋顶悬下的巨型玻璃罩其实是音箱,也就是说,在这件玻璃屋内,你可以带自己的音乐来就餐,营造属于自己的小空间而彼此不会互相打扰。相对正式的晚餐区,色调是黑、略深的木色,以及少许金色。内里的两间包房,分别用白色与深木色为装饰,给人以不同色调的冲击感。

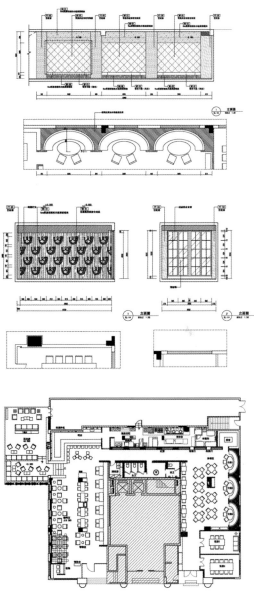

Chowhaus is surrounded by green trees and occupies a small area. Its wooden appearance looks like natural and low-key.

The open space of a restaurant is generally divided into four areas, where the right and middle parts are set with seats suitable for lunch while sofas and small round tables at the left side are arranged for drinking; generally speaking, that is the arrangement of the open space in a restaurant. However, Chowhaus is quite different. Walking inside along the left side, you will see a glass house. There lie many plants and a fireplace-installed bookshelf where old leather trunks and tourist handbooks collected from various places are displayed. The wooden table in primary color is surrounded by sofa in cream and grey colors. With sufficient natural light, the house looks more like an independent coffee room.

Three huge glass covers suspended under the roof are actually sound boxes. So, in this glass house, you can bring music you like to have meals, with your own space made here and without disturbance to each other. The dinner area is relatively formal, where black color and a little dark wood color together with some golden color are designed as the hue. Two compartments inside are decorated respectively with white color and dark wood color, bringing impact sense of different hues.

HASSELL

HASSELL

HASSELL是一家国际化的专业设计公司,拥有14间事务所,分布于澳大利亚、中国、东南亚和英国。公司有超过900名员工,业务经营已有70年历史,涉及全球多类市场。

HASSELL作为一家跨专业设计公司,拥有多领域专业实力,包括建筑设计、室内设计、景观设计和规划设计,公司注重可持续发展理念与设计实践的结合,兼备卓越的城市设计能力。

作为一家私营合伙公司,HASSELL的每一家事务所都具备了充分的灵活性和自主性,可同时为本地客户和国际客户提供服务,公司的综合资源和集体专业经验则为各地工作室提供了无可比拟的优势。

2012年世界建筑设计(BD World Architecture)公布的世界建筑设计公司100强中,HASSELL在澳大利亚设计事务所中排名第一位。HASSELL已获得超过650项设计大奖。

Cuisine Cuisine Beijing

北京国锦轩

设计单位:HASSELL
设 计 师:David Tsui 徐敏聪
项目地点:北京
建筑面积:2700 m²
设计时间:2011
摄 影 师:马未啸

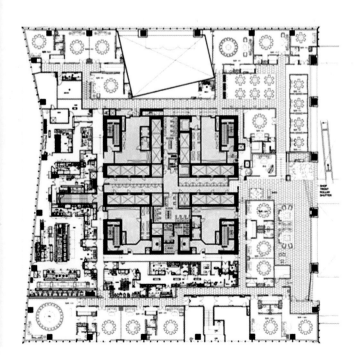

北京国锦轩位于北京CBD区的国际金融中心。这是一家正宗顶级粤港精膳食府,是香港米其林二星级粤膳食府国金轩在中国大陆地区的旗舰餐厅。

北京国锦轩位于北京市中心中的中心,故是以"中中之中"作为设计概念。这里面积达3000 m²,设有17间贵宾包房,可容纳约240位宾客。北京国锦轩以三种特别能够代表现代北京的颜色来渲染当地的特色历史文化层面。代表王府文化的金色,官府文化的红色,以及象征胡同里多年流传着的北京民间生活文化的灰色调。设计师凭借对中华文化敏感,把这三种极具视觉冲击力的色调融合起来,再运用这三种纯色各自的象征意义,来凸显出北京国锦轩作为一个在现代北京的中餐厅自身的文化传承亮点。设计师进而以不同的材料和质感,深化且突出了金色、红色和灰色各自不同的象征与特点。例如雍容华贵的金黄色,是贵宾房天花板上的巨型刺绣画作;喜滋滋红彤彤的是包厢里一整面的皮革墙;以及用来演绎城墙和传统四合院的灰色调景观等。三大主色调的巧妙运用,让人一走进北京国锦轩就能够感受到浓郁的京韵文化氛围。

此外，餐厅的规划也让宾客可享有更多用餐选择。例如，宾客可以选择在设计得像一个偌大庭院的开敞式大厅内用餐；或者可以选择在凉亭垂帘下的半开放空间里与朋友谈天叙旧，又或者选择在具私密性的贵宾套房里宴客。

其中，最具特色的是一组设计灵感来自于北京四合院中间公共活动区域的"半开放式用餐区"。墙面上，金色的叶子、红色的花卉和灰色的小鸟，用来点缀与装饰。在室内打造出京城大户人家后院般舒服自在的用餐环境。

Being located at the World Financial Center of Beijing's CBD, Cuisine Cuisine Beijing serves top authentic Cantonese cuisine. It's the flagship restaurant of Hong Kong's famous 2-Star Michelin Cantonese cuisine restaurant.

With "being at the center of the center downtown" as the design concept, Cuisine Cuisine Beijing is located at the center of center Beijing city. Covering an area of 3000 square meters, Cuisine Cuisine Beijing features 17 private dining rooms with a total seating capacity of 240 people. Cuisine Cuisine Beijing uses three colors that can specifically represent modern Beijing to demonstrate local specific historical culture. The three colors are gold that symbolizes the culture and history of the Forbidden City, red denoting culture of officials and grey that embodies Beijing's folk life and culture in Hutongs. Being sensitive to culture of China, the designer blends together the three hues which are with great visual impacts and applies respective symbolic meanings of the three pure colors to highlight Cuisine Cuisine Beijing's highlights on cultural inheritance as a Chinese restaurant in modern Beijing. The designer further uses different materials and textures to deepen and highlight respective symbols and features of gold, red and grey. For example, both dignified and graceful gold is used in the massive embroidery

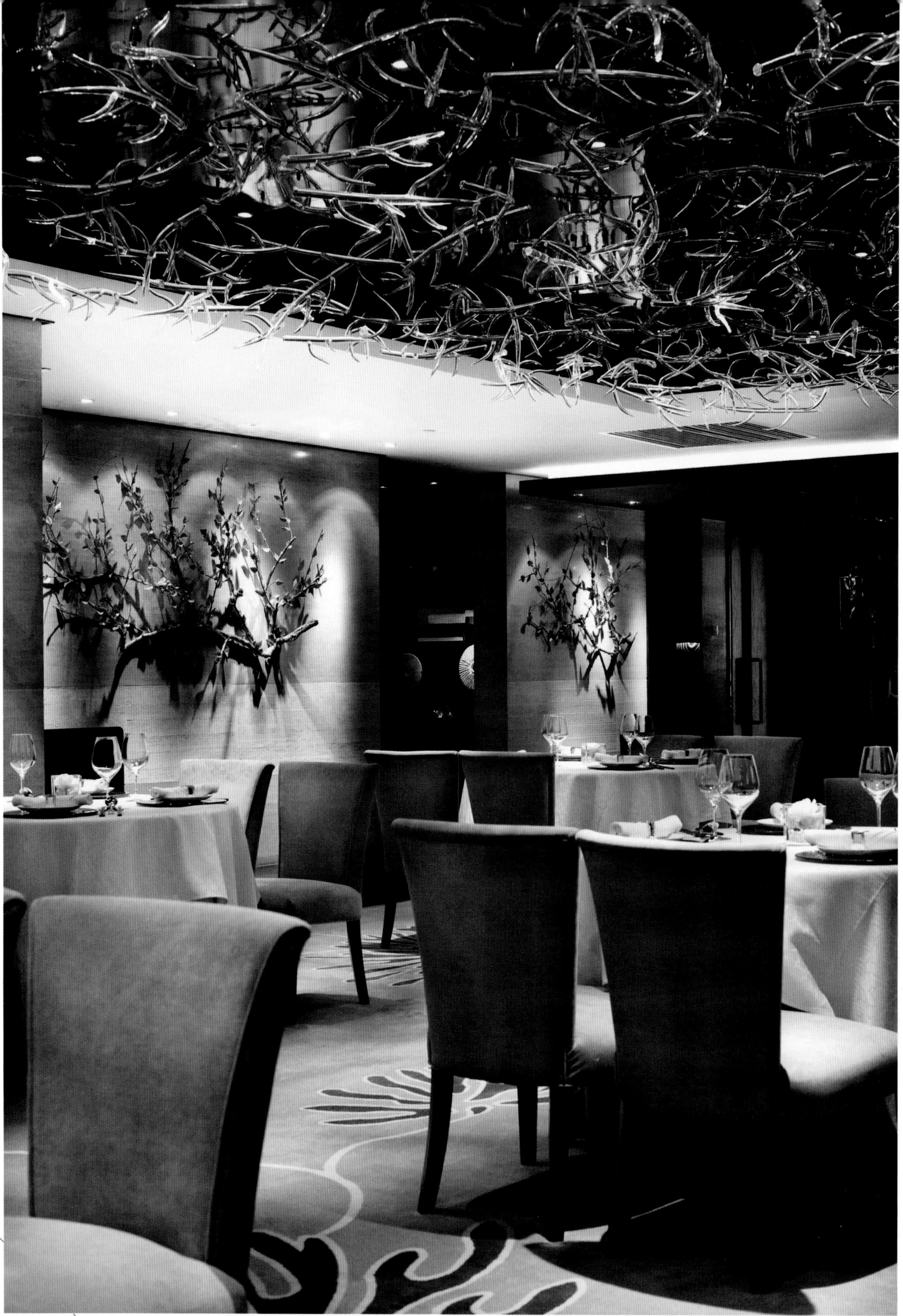

painting on the ceiling of VIP rooms; happy and prosperous red is the color of the whole leather wall of compartments and grey is used to demonstrate defensive walls and traditional quadrangle dwellings, etc. Delicate use of the three main hues brings people strong cultural feeling of Beijing as soon as they step into Cuisine Cuisine Beijing.

In addition, the planning of the restaurant enables guests to enjoy more flexible dining options. For example, guests can choose to dine in a designed opening hall which is like a huge courtyard, or chat and talk about the old days with friends in semi-open space with curtain in a pavilion or select private VIP suite to entertain guests.

Among them the most unique is a "semi-open dining area". The design inspiration of it is from the public activity area in the middle of Beijing's quadrangle courtyard. Golden leaves, red flowers and grey birds on the wall are used for embellishment and decoration and they create a comfortable and homelike dining environment similar to backyard of Beijing's rich family.

Concrete

荷兰著名设计公司,Concrete 的整个团队约有35位职业人士:他们是视觉市场商人、室内设计师、平面设计师和跨学科团队中参与项目建设的建筑师。

mazzo amsterdam
Mazzo餐厅

设计单位:concrete
设计团队:Rob Wagemans, Ulrike Lehner, Marc Brummelhuis, Sofie Ruytenberg, Femke Zumbrink, Rrik Van Dillen
项目地点:荷兰
建筑面积:400 m²
摄 影 师:Ewout Huibers

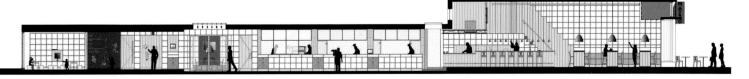

Mazzo是一家意大利风情的餐厅。这是一栋典型的阿姆斯特丹建筑：深而狭窄的空间，天花板的高度不尽相同，地面空间也被分成了好几块。地段也非常适合开餐馆。餐厅的第一部分坐落在玫瑰运河边，这里的天花板高达5 m，入口朝着运河。这部分最适合做成吧台，客人们可以点一杯速溶咖啡或者在这里喝一杯。灯光采用GUBI的Bestlite系列，营造了一种令人感到亲切的氛围。靠墙放了一张9 m长的低靠背大沙发。墙上和吧台上方的挂灯用了大小不同的两种型号，创造了一种居家客厅的气氛。

　　餐厅的第二部分矮了许多，但是宽了一倍。几根原石石柱将这块区域分成了2个部分。厨房是整个餐厅里最暗的地方，它对面就是用餐区。餐桌的安排简单灵活，也都是餐饮店常用的桌子。餐桌的摆式一般都是2人桌或者8人桌，每张桌子的位置都能让人一眼看到厨房里忙碌的大厨们。5组由MOOOI设计的Dear Ingo灯散发出柔和的灯光，照亮了整个区域。这些现代时尚的枝形吊灯给每一张独立的餐桌营造出愉快而亲切的氛围。墙上挂着4幅年龄各异的肖像画，客人在用餐的同时，能体会到意大利家庭式的风情。厨房的隔壁是一间舒适的会议室，里面有一个壁炉，关上门后，客人就能享受到一个舒适隐私的空间。这里适合用来举行商业会议或者家庭聚餐。

　　餐厅的第三部分向东直达Bloemstraat街。这里有一条小巷被改造成了一间密室，白天孩子们可以在家长们的监护下在里面玩耍。到了晚上，玩具会被收到一条黑白相间的窗帘后面。客人们可以靠着一条11 m长的靠背沙发上享用美酒，和朋友们聊天。Dear Ingo的灯光又一次在这里创造了家庭氛围。

　　作为空间融合多样性以及餐厅自然布局的连接，一个巨大的木质橱柜贯穿了整个餐厅，将所有的空间连接起来，进行统一分配。这个坚实的巨大松木橱柜不仅作为产品陈列柜，还是通往阁楼的楼梯。这里还能作为吧台后方、食物配送点、会议室与餐厅之间的墙、衣帽间、休息室的通道以及儿童玩具的储藏室。

　　所有这些粗犷真实的设计由五种原料构成：水泥，整齐的砖块，石头，松木以及原钢。前三者取自建筑本身，原钢和松木是新加上去的。

　　大橱柜的窗架以及餐厅前方的阁楼完全由原钢搭建，这些钢制的梁柱暴露在外。阁楼用未经处理的原钢网围造。这种直接采用原始材料的设计是为了能使客人不被装潢分心，餐厅更专注的是料理本身。

　　在设计标志时如何融合餐厅的历史和名字掀起了热烈讨论。Mazzo标志是由LED灯在原钢框架上排列组成，让人联想到餐厅的前身是迪斯科舞厅。整个标志如同一顶闪亮王冠悬挂在吧台上方，就算是路过汽车的车灯或者自行车车灯都不能淹没它的灯光。

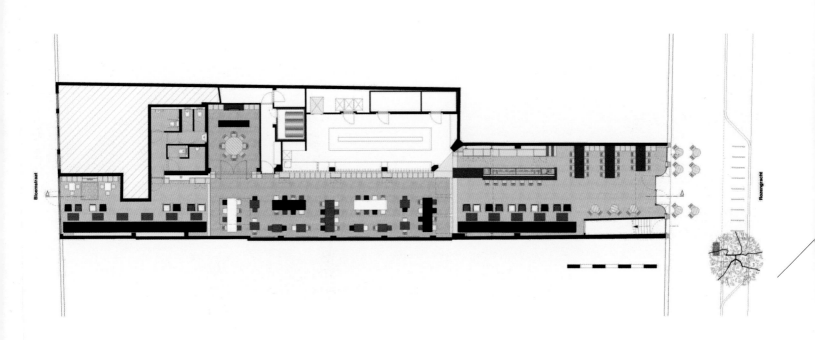

Mazzo is an Italian restaurant. The building is typical in Amsterdam: narrow and very deep spaces fused together with different floor and ceiling levels, automatically providing a natural position for the restaurant.

The first part, situated on the Rozengracht, has a five-meter high ceiling and faces directed outwards. The perfect area for a bar, where guests can order a fast espresso or have a drink at the high bar tables. Lighting from the Bestlite series by GUBI creates the intimacy in this space. The wall lights above the nine-meter chesterfield couch and the suspension lights above the bar and bar tables in xs and XL sizes create a living room ambiance.

The second part is obviously lower but twice as wide. A few original stone columns divide this part into two zones. The darkest part of the entire building is perfect to house the kitchen. The seating area of the restaurant is opposite to the kitchen with a simple but flexible, trattoria-like table arrangement. A variation of tables for two or eight are placed and every table has a great view of the chefs in the kitchen. Five Dear Ingo lights by MOOOI comfortably light up the seating area. Independent of the table arrangement, these modern chandeliers give a pleasant and intimate light. Four portraits of people in different age categories provide an Italian family-feeling of which the guests can be a part during dinner. For a real family dinner or a business meeting guests can use the boardroom. A cosy room with a TV screen and fireplace, next to the kitchen, that can be closed off if privacy is preferred.

The third part is orientated towards Bloemstraat. A narrowing in the building creates the small backroom where kids can play during the day under the watch

of their parents. At night the toys can disappear in the cabinet behind a black-and-white blocked curtain and guests can retreat into the eleven-meter long chesterfield couch to enjoy a good glass of wine and have a nice conversation. Again the Bestlite wall lights by GUBI create a home-like feeling.

The diversity of the fused spaces and the natural restaurant layout need a connecting element: a huge wooden cupboard across the whole restaurant linking all the spaces to each other and organising them at the same time. The cupboard, created out of solid pine wood for storage and display of the products, becomes the stairs to the mezzanine floor, the back bar, point of distribution for the food, the transparent division between the boardroom and restaurant, the wardrobe, the access to the restrooms and the storage for the kids toys.

Five materials determine the ambiance of the raw and honest interior design: power floated concrete, chipped brickwork, stone, pinewood and raw steel. The first three materials are part of the shell of the building; all the new materials are steel and wood.

The window frames in the cupboard and the mezzanine floor in the front of the restaurant are completely made of raw steel. The steel beams and columns are exposed and the extra floor is provided with an untreated expanded metal mesh. The use of honest and simple materials doesn't distract the guests' eye and underlines the fact that the restaurant focuses on the food.

Embracing the past of the building and the name Mazzo by creating a logo that is a controversial part of the venue. The five letters are made of raw steel and filled with classic amusement lights, referring to the disco days of Mazzo. The light object crowns above the bar and will be noticed even in a flash by the cars and cyclists passing by.

MARCELO JOULIA

1958年出生于阿根廷
法国注册建筑设计师和城市规划师
纳索巴黎 & 纳索上海（naco architectures）设计总监
Architect D.P.L.G. 建筑设计师

Adam Fang 方钦正

毕业于英国曼彻斯特大学建筑系
现任法国纳索建筑事务所设计总监/建筑师/合伙人

Unico + Colagreco

唯壹餐厅

设计单位：纳索建筑设计事务所
设 计 师：Marcelo Joulia，Adam Fang
项目地点：外滩三号
建筑面积：1100 m²
设计时间：2012.06
摄 影 师：Derryck Menere

"唯壹"提供别具拉丁风情的用餐品酒环境。独特的室内空间来源于建筑设计师的精心设计,并通过独树一帜的时尚设计元素为宾客创造前所未有用餐体验。其中,这些元素包括:设计感强烈,摆放错落有致的古典家具,全上海独一无二的巨型冰雕灯架。此外,异域风情的装饰灯光,摩登复古的地毯,一只叫做COCO的白色鹦鹉作为餐厅的吉祥物当仁不让成为瞩目亮点。

"唯壹"也将与NACO建筑事务所一同建立一个为当代艺术家提供展示他们创新杰作充满活力的平台。从现场表演到由专门的音乐制作人悉心创作精选的音乐,所有的艺术项目都将成为餐厅环境里绝妙的节拍。

"唯壹"Tapas酒廊位于上海外滩三号这个为歌颂"当今的生活艺术"而生的历史建筑二楼。整个空间是由连续的开放式厨房以及令人瞩目的大理石吧台组成。而颜色,造型以及具有独特纹理的装饰材料则配合并呼应着以主要空间为主角的整体氛围。

在"唯壹"内一座绿色的透明玻璃房,里面设置了绿色植物配合180度观赏外滩江景和江畔景观,拥有截然不同的视觉体验。

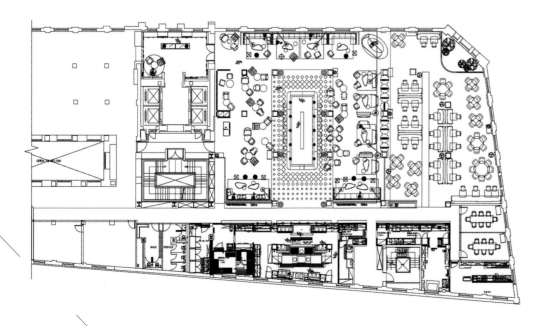

Enlarging Latin culture in a festive space, UNICO interior design is the result of a mature reflexion, illustrated by the most iconic design elements: a patchwork of vintage furniture, a unique ice sculpture lighting suspension, without mentioning decorative lighting, carpets and even a white parrot as mascot called *Coco* …

UNICO will also become a dynamic platform for contemporary artists to show their creations within Naco Architectures design. From live performances to sophisticated sound selection, the artistic program will give a fantastic tempo to the venue.

UNICO tapas lounge is located at Bund 3 building, a place full of history, celebrating *today contemporary art of living*. UNICO space is organized around the endless open kitchen and impressive central marble bar. Using refined raw materials, the interior design is playing with colors, shapes and textures in an unexpected yet consistent way.

At the waterfront, over the Bund promenade, a green patio, veritable jungle echoing the continent rich nature is creating an extra visual experience.

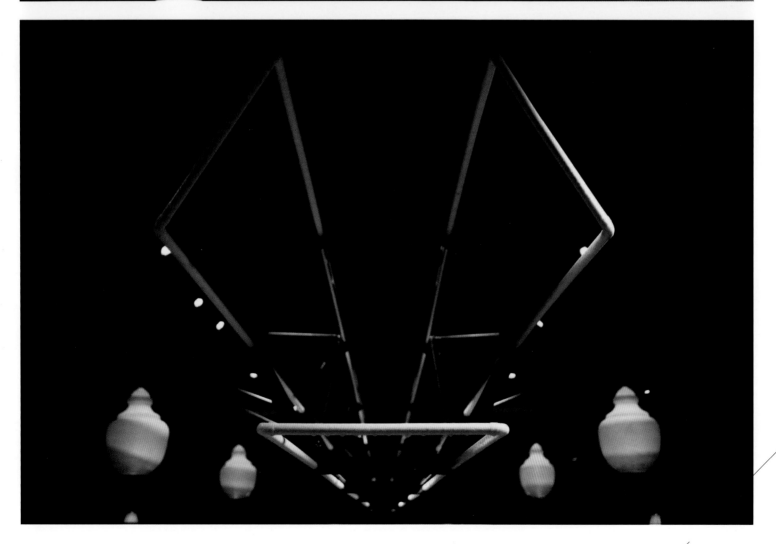

甘泰来

齐物设计事业有限公司总监
交通大学建筑研究所兼任讲师
哥伦比亚大学建筑硕士（荣誉获奖毕业）
康奈尔大学建筑与都市设计研究所研究
东海大学建筑学士

Sabatini-SH 餐厅

设计单位：齐物设计事业有限公司
设 计 师：甘泰来
参与设计：高泉瑜、张芃欣、黄盈华
项目地点：上海
建筑面积：室内 430 m²、室外 94 m²
主要材料：洞石石材、沙比利木、琥珀玻璃、
　　　　　玛莱漆、意大利复古砖
摄 影 师：卢震宇

基地位于商业大楼的地面层,设计师以"盒"状的设计语言,表现空间层层布局的趣味性。

入口处以内凹的方式界定出一座接待门厅,穿越门厅来到了吧台、用餐区、包厢或露台区。特别借由既有的梁柱增建成拱廊造型,让拱柱优雅的轮廓划分出前后段的用餐区域;而层层的门框、拱廊也区分了餐厅的动线转折与座区安排,将各餐桌距离加大,满足交谈隐私以及侍酒与桌边服务所需的空间。考量该区消费客层多为商务人士,设计师除了在入口处规划酒吧作为餐前交谊的场所,更在露台区安排沙发座让宾客能在此轻松交谊。

此外,餐厅特别规划一座VIP包厢,包厢以镶嵌玻璃围砌而成,在水纹玻璃的中介下让内外视景产生朦胧效果,而达到了包厢内部隐私,长型而挑高的空间在玻璃的高透光性而不产生窄迫感。厢房内,可透过一道暗门弹性连结至内部厨房,让贵宾能亲见主厨私房菜的创作过程,包厢也可以再进行场域划分,满足不同人数的餐宴需求。

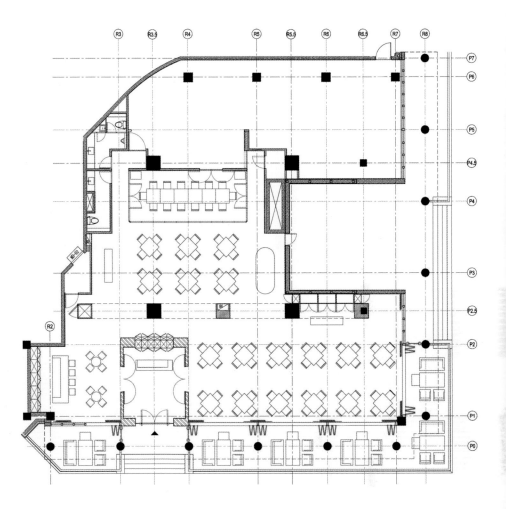

The base is located in the foundation floor of commercial building. The designer uses the designing language of "box" shape to express the interesting of layer upon layer layout.

Within the entrance concave way to define a reception hall, you can across the hall come to the bar, dinner area, box, and balcony area. Special by the existing beam-column built arcade modeling, the counter fort elegant profile divided the dinner area into front and back area. Additionally, Layers of door case and arcade also distinguished between the dining room of dynamic line turning and sitting area arrangement, increased the distance between tables to meet the needs of talking privately, servicing wine and needing space of service beside the table. Considering the customers in that area are businessmen, the designer not only plans the bar at the entrance as the trading area before the dinner, but also arranges the sofa at the balcony area to let the customs communicate relaxed.

In addition, there is a special VIP-case in the restaurant within mosaic glass, in the water cut glass intermediary to internal and external visual produce hazy effect to achieve the compartment internal privacy, long and high space in glass high-translucent do not produce narrow forced feeling. In the wing, through a fly elastic links to internal kitchen, guests can see the chief cook's private kitchens creation process, box can also further divide the field, meet different number of people's meal demands.

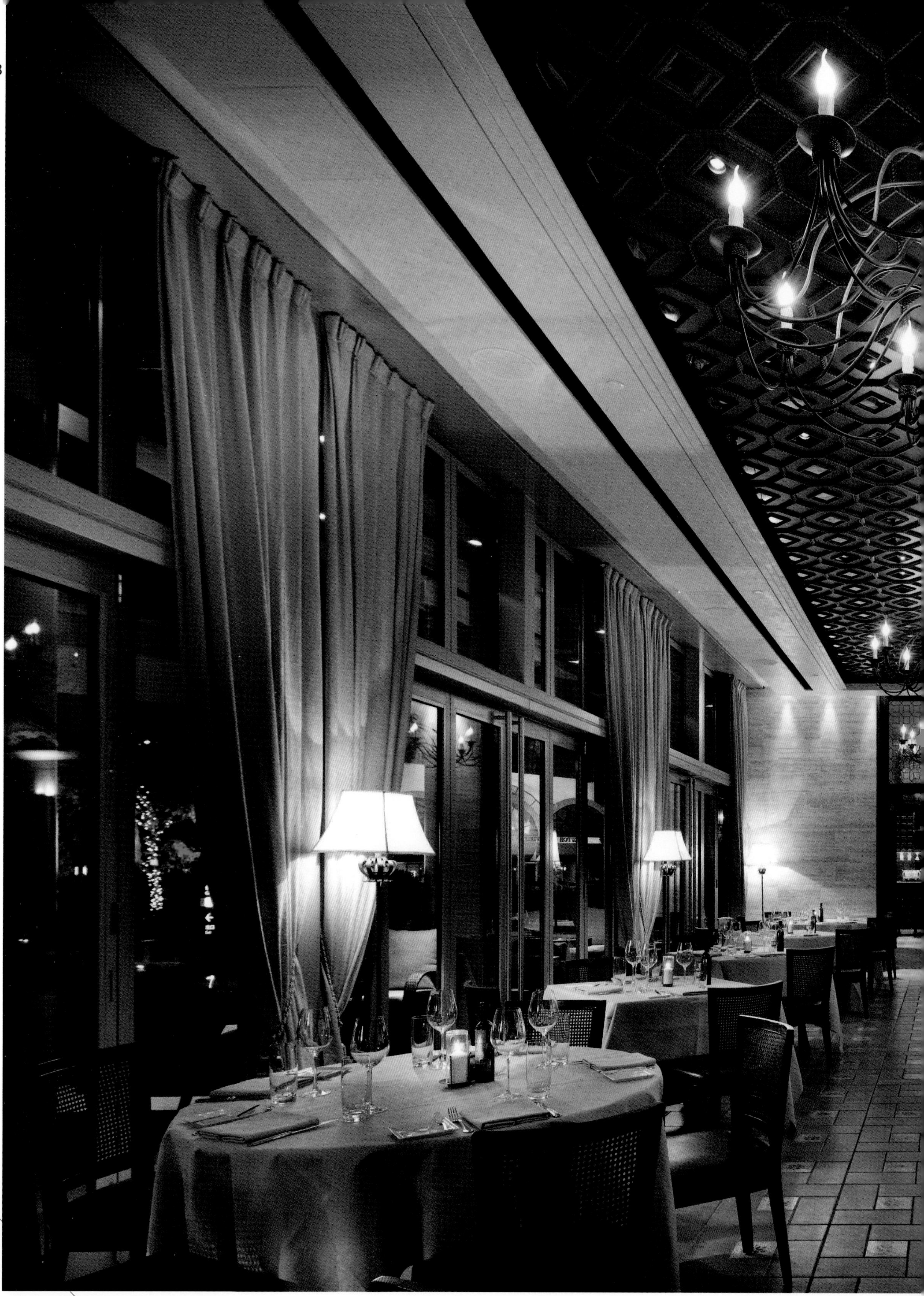

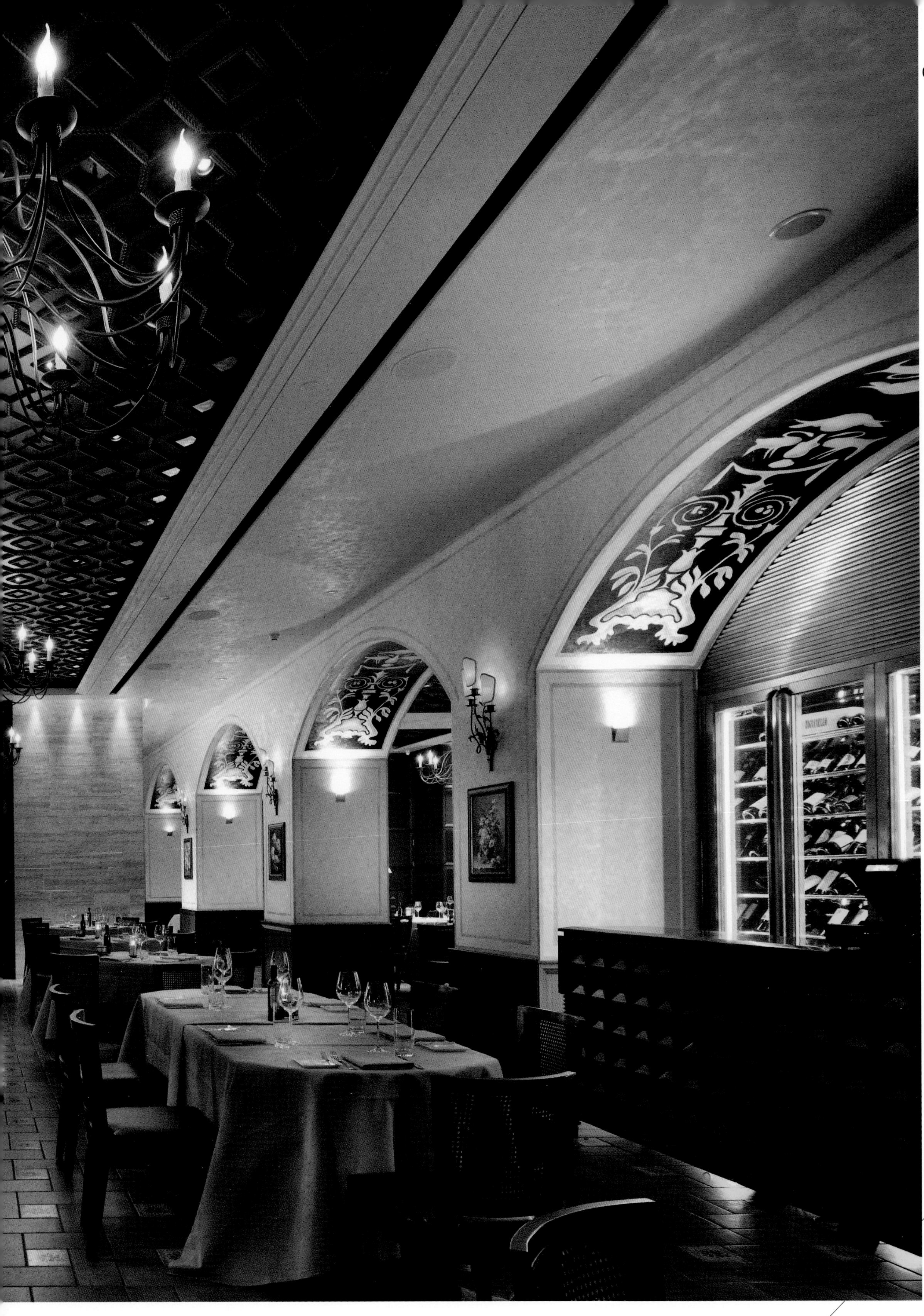

陈德坚 (Kinney Chan)
德坚设计创办人
曾任香港设计中心董事
现担任香港室内设计协会会长

Sesame Japanese Restaurant
Sesame 日式餐厅

设计单位：德坚设计
设 计 师：陈德坚
建筑面积：330 m²
摄 影 师：Trio Photography,
　　　　　Michael Perini

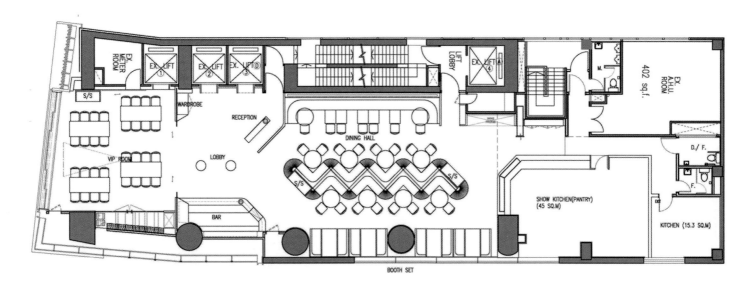

Sesame 是一所位于中心地带的高档日式餐厅,它的设计极富当代艺术感,显得简洁而安静。餐厅标志创造性地使用了一系列椭圆形图案,象征着芝麻,与餐厅的名字相呼应。

为了打造多变的自然风格,设计团队使用了大量的自然素材,结实的木材、光滑的石板、粗犷的岩石以及细致的沙子等随处可见。

餐厅的设计十分用心,整体包括一个由卡座构成的用餐区,一间VIP室,一个寿司吧台以及一个独特的开放式厨房。

餐厅的服务台是特别定制的抛光木结构,水槽则是竹结构,这些灵感都来自日本的传统元素。大厅的天花板用了大量枫木条板。整个餐厅的的设计极富韵律感,舒适宜人。

VIP室的入口由木头构成,看起来像个山洞。天花板上装饰着椭圆型的灯泡,像极了芝麻。窗口的形状极不规则,象征着山。它的墙也是一大特色,上面展放着各种日本名酒,使人感觉身处一座名酒博物馆。这样的设计不仅给客人带来了视觉上的愉悦,更让餐厅显得极为高档。

Sesame Japanese Restaurant is an upscale dining place located in the heart of Central. This Japanese-inspired design has a contemporary interior design aesthetic and simple quietude. To bring out the significance of the restaurant title, an inspiring logo with numbers of oval pattern that represents sesame seeds is created.

Attempting to bring different ambience of nature into this environment, designer makes frequent use of natural elements such as fine woods, slate, rocks and sand.

The restaurant is efficiently planned out, the layout includes a main dining area with booth and table seating, a VIP room and a distinctive show kitchen and sushi bar.

Taking inspiration from the tradition Japanese elements, a custom millwork-like reception counter and a bamboo shaped water sink is created. On the other hand, the ceiling in the main area is decorated by the endless amount of vertical maple slats that line the ceiling into a roof shelter. It generates a sense of rhythm and a perfectly cozy atmosphere at Sesame.

Passing through the cave-like wooden porch is the VIP room. The room design highlights the ceiling panel with oval shaped lightings that resemble sesame seeds. Irregular pattern on the window frame is the shape of a mountain. Another dramatic feature is the wine display along the wall, and it is a walk-in library of Japanese fine wines. This design not only generates visual pleasure to the visitors, but also makes the restaurant remarkably luxurious.

张永和

建筑师,建筑教育家,非常建筑工作室主持建筑师;

美国注册建筑师;

美国伯克利加利福尼亚大学建筑系硕士;

北京大学建筑学研究中心负责人、教授;

2002年获美国哈佛大学设计研究院丹下健三教授教席;

2005年就任美国麻省理工学院(MIT)建筑系系主任。

Beijing Kingsjoy
北京京兆尹

设计单位:非常建筑工作室
设 计 师:张永和
创意构思:京兆尹
项目地点:北京市东城区五道营胡同2号
　　　　　(雍和宫大殿西侧)
建筑面积:2210 m²
主要材料:砖、瓦、织物
设计时间:2011.08
摄 影 师:舒赫

京兆尹总营业面积2210 m²,可容纳240人同时就餐及举办各类活动。院落环境舒适、风雅,现代科技负离子喷雾迎宾,洗涤一身尘嚣,处于闹市,犹如置身于自然森林中;走进中庭观景院,春揽百花秋邀月、夏沐凉风冬嬉雪;四周金砖铺地、瓦当窗花,人亦融入环境中成为一道风景。宾客在此体验管家式服务、沙龙雅集品茗区、无触碰式卫生间、墙壁上、蚕丝吊灯上处处尽显细节之美,身心也随之轻安。

Total business area of king Join is 2210 m² capable of 240 people repast and holding all kinds of activities at the same time. Amenity and elegant courtyard, greeting guests with anion atomizing by modern science and technology, washing dirt over the whole body, as if place yourself in natural forest though in downtown area. You may accompany by all sorts of flowers in the spring, avoid summer heat in the summer, invite bright moon to enjoy the night in the autumn and play with snow in the winter in the viewing courtyard. Golden bricks cover the ground all around here; eaves tiles take the role of paper-cut for window decoration and people integrate with the environment and form another great landscape. Guests may experience housekeeper service here, while beauty of details is everywhere such as in elegance gathering salon, in tea appreciation area, in no-touch style restroom, on the walls and on the natural silk droplight… Being in the elaborately designed environment, the mood will be comforted accordingly.

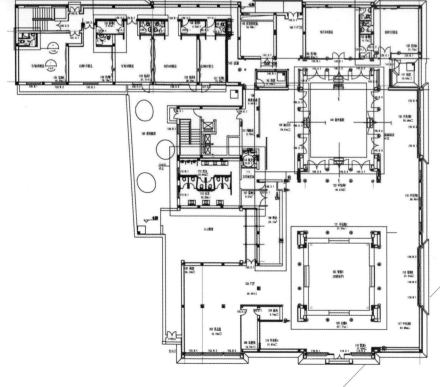

甘泰来

齐物设计事业有限公司总监
交通大学建筑研究所兼任讲师
哥伦比亚大学建筑硕士（荣誉获奖毕业）
康奈尔大学建筑与都市设计研究所研究
东海大学建筑学士

纱舞缡 ATT 4 FUN 店

设计单位：齐物设计事业有限公司
设 计 师：甘泰来
参与设计：何智渊、刘煜铃、郑明安
项目地点：台北市信义区 ATT 4 FUN
建筑面积：281.7 m²
主要材料：镜面不锈钢、明镜、茶镜、茶玻、南方松实木染色、柚木木皮、黑色砌石素面石英砖、金属砖
摄 影 师：卢震宇

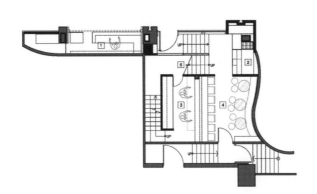

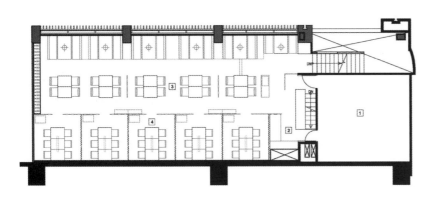

ATT 4 FUN的纱舞缡餐厅位于台北市精华地段里的信义计画区,挑高两层楼的入口接待区,充满现代设计的空间意象。部分一楼空间除了室内寿司吧台和座位区,还加上提供外卖的黑轮吧台区;楼梯入口利用镜面拉伸橱窗端景的延展效果,天花板矩形不锈钢刻意高低垂吊下来,营造吊灯概念融入空间,木块墙掺杂反射面于其中的凹凸之间;寿司吧则选择实木与石材的搭配。二楼,包厢区则利用茶色玻璃划分出半个开放场域,多为开放的状态下,可透过帘子来区隔用餐空间,依照消费人数的需求调整座位安排。

　　开放式天花板设计,利用骨料钉制成格子状,于交接处安装矩形镜面不锈钢,产生一个半穿透感的天花板结构,透过不锈钢反射面,模糊掉原本较矮的天花板高度,加上天花板重点的投射照明,其不锈钢镜面巧妙反映着室内光源,不同角度都形成有趣的空间亮点,用餐时让人感到不刺眼。立面墙借由方格实木分别镶嵌在玻璃面上,展现不同层次的丰富面貌,转移原有矮小尺度的焦点,营造模糊视觉界线的空间主题。

Saburi restaurant of ATT 4 Fun, which has an about two - storey high entrance reception area and is full of space image of modern design, is located in the best section of Sinyi planning district in Taipei city. Besides indoor Sushi bar and seating area, a hotchpotch bar which offers take-away exists in part of the first floor space; the foot of the staircase, which stretch the extension effect of the end view of show window using mirror, the rectangular stainless steel of ceiling deliberately hangs down high-fall to construct drop-light concept into space, block wall and reflector mix in the concave-convex; and Sushi bar chooses to collocate solid wood and stone together.

The open ceiling design, which uses aggregate nail to make into the grid and installs rectangular mirror stainless steel in the connection, to produce a ceiling structure with half penetration sense, plus the projection lighting of the ceiling focus, and the stainless steel mirror which cleverly reflects the interior light source, the different angles all form the interesting space window, these all make people feel not dazzling when they have dinner. The gables are inlayed in the glass by the grid solid wood respectively, which exhibit the rich appearance of different levels, and transfer the original small scale focus in order to construct a space theme of fuzzy the line of sight. In the second floor, balcony area is divided into half an open field using the solar glass, under the open condition, the restaurant can separate have dinner space through curtain and they can adjust seat arrangement according to the requirement of consumption number.

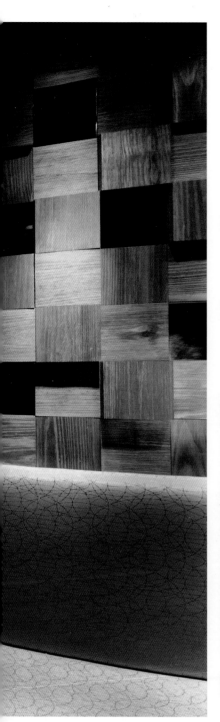

利旭恒
出生于中国台湾
英国伦敦艺术大学荣誉学士
古鲁奇公司设计总监

YAKINKUMASTER
Japanese barbecue Restaurant
烧肉达人
日式烧肉店

设计单位：古鲁奇建筑咨询（北京）有限公司
设 计 师：利旭恒
参与设计：赵爽 季雯
项目地点：上海
建筑面积：300 m²
竣工时间：2011.11
摄 影 师：孙翔宇

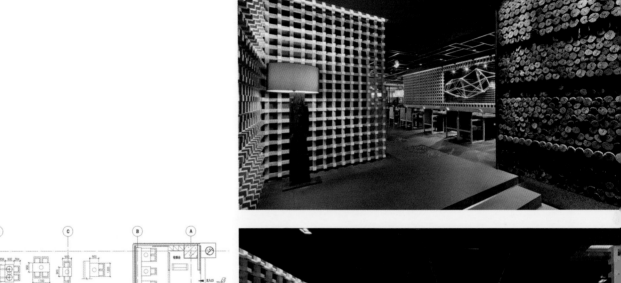

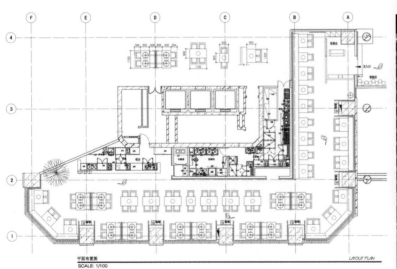

YAKINIKU MASTER烧肉达人日式烧肉店位于上海天钥桥路上。品牌创立人期望能将日本禅意与中国江南水乡的概念移植到上海,让宾客在舒适优雅的空间里享用美食同时,感受到文化的氛围。设计师利旭恒运用现代的手法演绎日本传统建筑的基本框架结构,大量的木框架朴实地表现建筑结构美学,另外用水墨方式呈现江南水乡中国建筑屋脊的曲线,曲线来自屋瓦,依着梁架迭层的加高,借此强调了中国建筑结构上的不可置信的简单和自然。

YAKINIKU MASTER Japanese barbecue restaurant in Shanghai Tianyaoqiao road, Brand owners aspire to Japanese Zen transplantation with the concept of south China's country style to Shanghai, so that guests enjoy the meal in a comfortable and elegant space at the same time feel the atmosphere of culture. Designer Lee Hsuheng use of modern techniques to interpretation of the basic frame structure of traditional Japanese architecture, a large number of simple wooden framework of the performance aesthetics of building structures, the other showing the curve of the south of Chinese building roof with ink, the curve from the roof tiles in accordance with the beam the rack layer heightening, thereby emphasizing the beauty of this curve in the Chinese architecture construction almost incredible simple and natural.。

陈彬

武汉理工大学艺术与设计学院副教授\硕士生导师
CAAN 中国美术家协会会员
CBDA 中国建筑装饰协会设计委员会委员
CIID 中国建筑学会室内设计分会会员
IAI 亚太建筑师与室内设计师联盟理事
ICIAD 国际室内建筑师与设计师理事会会员
大木设计中国理事会常务理事
后象设计师事务所创始人\设计主持

Design Description of the Inked Orchid Pavilion

水墨兰亭

设计单位：后象设计师事务所
设 计 师：陈彬
项目地点：武汉
建筑面积：649 m²
竣工时间：2011.12
摄 影 师：吴辉

这一餐饮空间是兰亭主题新的尝试,设计的焦点集中在"水墨"一词中,设计师选用拉丝黑钢,白色云石和染色木作,将主色调严格地控制在"黑、白、灰"范围内,决不逾越雷池一步,纯白质地装饰感极强的浅浮雕,将中国传统文化中诗、书、画和古玩以极现代的平面手法表现出来,是空间中的最大看点。最后在完成的空间中选用高纯度的青绿布面的家具,既意外,又合理,流露出设计师对强调入世的中国文人画风的新解。

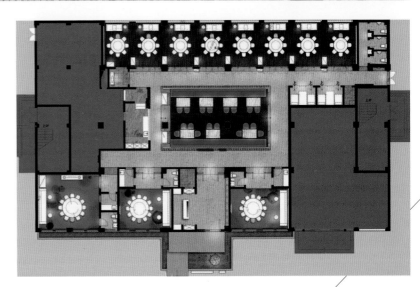

This dining space shows another attempt of the theme of Orchid-pavilion with a design focusing on the word "inked". The designer utilizes the wiredrawing black steel, the white marble and the dyed carpentry, and strictly limits the dominating colors to the range of "black, white and grey" without any overstepping of the limit. The pure white bas-relief with a strong sense of decoration demonstrates the poetry, the calligraphy, the painting and the curiosity in the ancient Chinese culture with modern plane technique, which forms the major feature of the space. Finally, the furniture in dark green of great purity is selected for the completed space, which is quite beyond expectation but meanwhile quite reasonable, revealing the designer's re-explanation to the Chinese scholars' secular painting style.

陆嵘

2002年4月至2006年12月 上海华东建筑设计研究院有限公司室内设计景观设计部主创设计师、主任助理；

2007年1月至2007年12月 上海现代建筑装饰环境设计研究院有限公司第三设计所常务副所长；

2008年1月至今 任上海HKG建筑咨询有限公司中方设计总监和主创设计师

设计的项目多次获得国家及省部级奖励，并获上海十大优秀青年室内设计师及2009年度中国首届中华文化人物称号

蔡鑫

现任上海HKG建筑设计咨询公司设计总监助理、主创设计师

曾就职于上海华东建筑设计研究院有限公司室内及景观设计部、上海建筑装饰环境设计研究院装饰三所

主要设计或参与设计了多个重要大型项目，包括大型办公楼、酒店、会所、宗教的多个领域

Walnew Club
Changguang wetland park, Wuxi

无锡长广溪
湿地公园蜗牛坊

设计单位：HKG GROUP
设 计 师：陆嵘、蔡鑫
参与设计：王文洁、吴振文
项目地点：无锡长广溪湿地公园缘溪道6号
建筑面积：2400 m²
主要材料：松木、红砖、铁锈金属、涂料、清水水泥地面、艺术羊毛毯、刨花木地板
竣工时间：2011.12
摄 影 师：刘其华

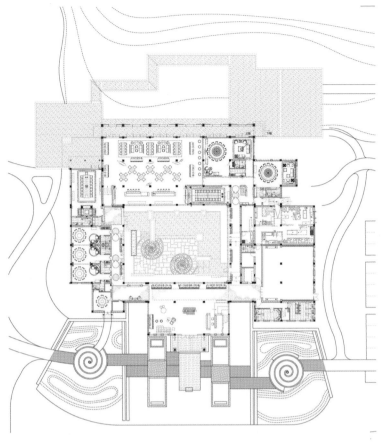

WALNEW CLUB——无锡首家"都市慢生活"的创意餐厅,将设计的创意、美食的享受与湿地的静谧、写意的自然环境融为一体。

结合建筑周围的生态环境,用自然质朴的材料与之相呼应。室内环境基础色调为中性偏冷,通过艺术装置的鲜丽色彩加以点缀,来打破平稳的节奏,从而提升视觉趣味性。家具灯具的设计均简约富有创意,细节之处体现来自大自然中的元素撷取。

室内整体造型线条流畅清晰,虚实有序。无论是墙上块面颜色的铺展,还是顶面的条形格栅造型搭接,始终以简练的几何关系诠释主题的定义:细腻、质朴与自然。

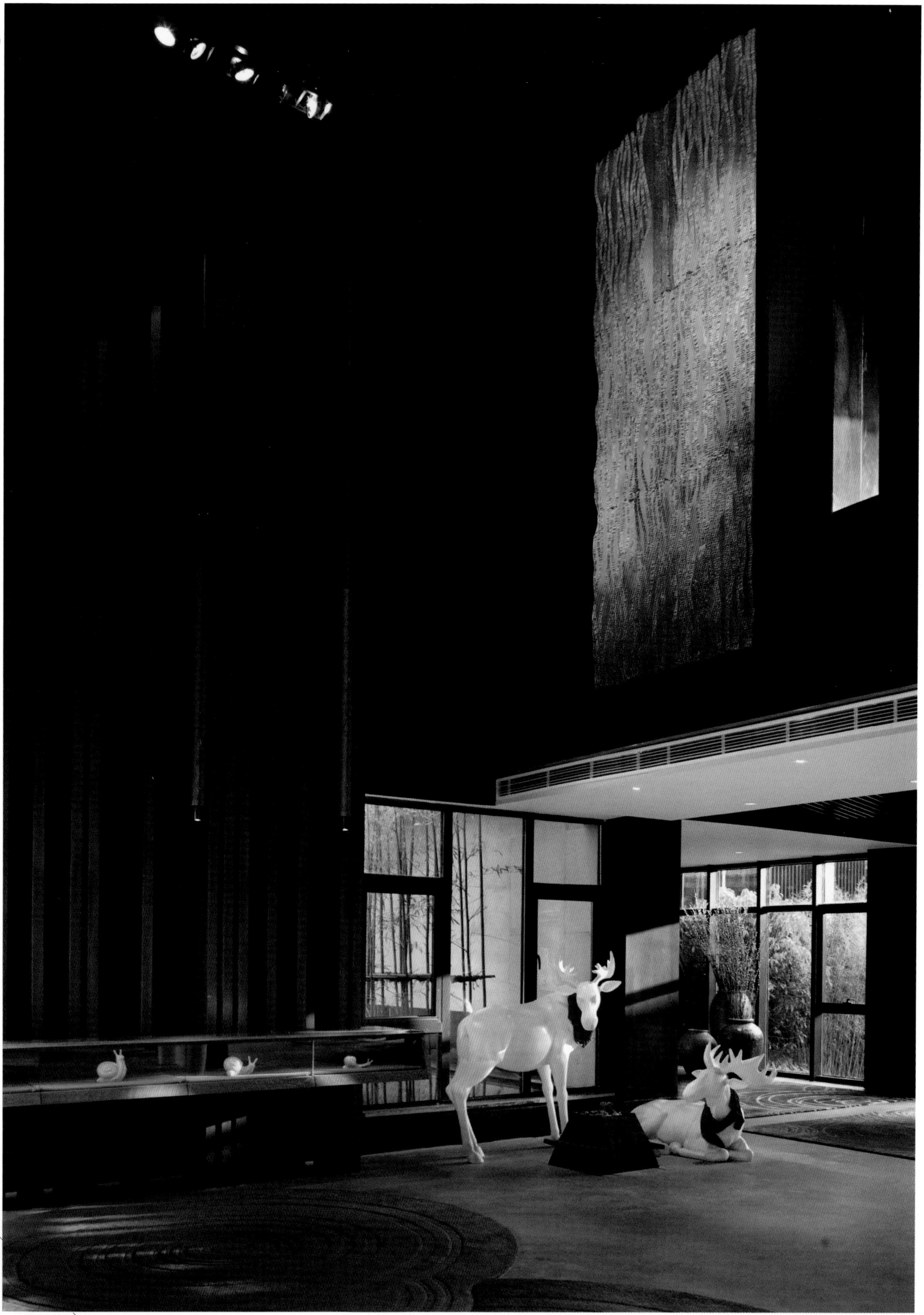

　　空间整体色彩以深浅两种灰色为主基调,大面积交织。红褐色黏土砖以醒目的颜色,通过特别的角度设计铺贴在部分空间的主要墙面,砖墙四周用自然锈斑表面的金属折成精致的条框收边。历史悠久的老木头映射出自然醇厚的颜色,与古铜色木饰面一并在空间内对话,交替出现,相互衬映。部分区域用大幅镜面单元框以角度错开的方式排列、间隔,点缀其中,折射出不同凡响的奇异空间。

　　地面的艺术地毯造型更加生动,以渗入湿地植物的造型和色彩元素加以抽象提炼,通过各种编织工艺来表现。使整个装饰色彩基调简约质朴却不失灵动。

　　蜗牛坊在打造特别的用餐环境及提供各类美食服务的同时,还是一个设计艺术的展示平台。在主要公共区域预留了展陈柜台和空间。定期举行不同主题的创意设计展品陈列,使宾客满足味蕾的同时更赏心悦目,真正感受到设计艺术与饮食文化的交融,获得多方位的全新感受。

WALNEW CLUB——the first creative restaurant of "slow life in urban area", which had integrated with design originality, enjoyment of delicious food, tranquil of wetland and enjoyable natural environment.

Considering the environment which is surrounded by buildings, with which use materials both of natural and simple to work concert. Base tone of indoor environment is neutral prejudiced cold, which had embellished by floridity color of artistic installation to break stable rhythm, therefore, which will increase visual enjoyment. Design both of furniture and lighting are creative and brief, and reflections of its details was captured from the nature.

Lines both of smooth and clearness on total look of indoor are false or true with order. Not only color spreading of pieces of wall but also lapping of grating modeling of bar-type on the top will always interpret definitions of themes by geometrical relationship as fine and smooth, simple as well as natural.

Whole color of spaces had taken 2 kinds of grey as main keynote, interweaving with large area. With striking color, brick of bronze clay was paved on main wall space in part spaces through design of special angle, all round of brick wall uses metals which surfaces is of natural rust to fold delicate frames which is close at the sides. Old wood of long history had re-

flected color both of natural and mellow as well as bronze timber facing, which had a dialogue in the space, both alternation with and setting off one another. There are substantial unit frames of mirror surfaces in part fields, which had arranged, spaced and embellished among them on stagger angles to reflect different spaces both of outstanding and strange.

Artistic modeling of carpets on the ground is more living, and abstract fining will base both on modeling of penetrating wetland plant and color elements, which will be reflected by crafts of various weaving to make color motif of whole decorations be both brief and plain without easy going.

During both building special catering environment and providing various delicious foods, snail workshop is also a platform for art show of design. There are both spaces and cases which are reserved in public areas. There are lines of creative design of different themes which is held regularly, which will make guests be better to hear and see while satisfying its taste bud, and really feel the integration both of design art and catering culture to gain brand new feeling of multi-aspect.

孙黎明

无锡上瑞元筑设计制作有限公司董事设计师
CIID中国建筑学会室内设计分会理事
CIID中国建筑学会室内设计分会第三十六（无锡）
专业委员会秘书长
江苏省室内设计学会理事
IFI 国际室内建筑师/设计师联盟会员
ICIAD 国际室内建筑师与设计师理事会理事
美国IAU艺术设计硕士

HBO Movie Theme Restaurant
HBO电影主题餐馆

设计单位：上瑞元筑设计制作有限公司
设 计 师：孙黎明
参与设计：陈凤磊、陈贝、陈浩
项目地点：无锡解放西路小尖上2号
主要材料：大理石、电镀不锈钢、绿可木、皮革打印图案、墙纸打印图案、白影木、水曲柳浮雕板、老木头
建筑面积：450 m²
竣工时间：2012.02

项目业态定位、风格诉求,都源于"HBO"业主陈先生的情结——欧美范儿、热爱经典电影。这是一个能让很多本埠人看得懂的欧美风尚餐厅、一种流溢"奥斯卡"情境的空间味道、一套"做实"的元素组合、一种夹裹怀旧的厚重不失清扬的空间气质,让目标群获得既精致又放松又饶有兴趣的身心体验。空间设计上,充分重视陈设的主表情作用,力求丰富、饱满,而装修则成为背景,为陈设精彩提供恰如其分的舞台——主材的持重含蓄、色彩的肌理自然、结构的大气朴茂。更多的是来自电影视界的细节,海报、胶片、老式电影机、影人肖像、唱片、"那个年代"的自行车等等,有应用要素的呈现,更有平面设计的巧思。

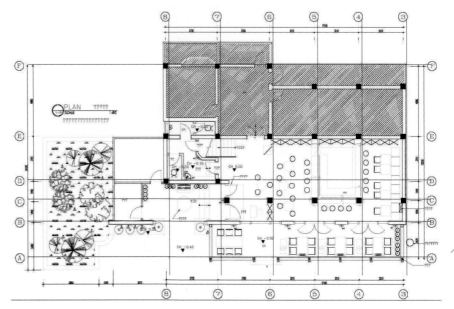

Both project orientation and style aspiration of the project are derived from the love knot of Mr. Chen, proprietor of "HBO". Construct a fashionable European and American style restaurant that is understood by most of local people, create a space with "Oscar" situation, integrate "real" elements and create a place with both nostalgic style and clean & light features to enable target groups acquire delicate, relaxing and interesting mind and body experiences. In the aspect of space design, the designer attaches importance to the main expression functions of furnishings and tries to achieve a feeling of abundantness and plumpness, while decoration is regarded as background to provide suitable platform to furnishings ---- Main materials show dignified and implicit features, while texture of color is natural and the structure is imposing and simple. However, more things are from details of movie industry: posters, films, outdated cin matographe, portraits of filmmakers, records, bicycles of "the age" and the like… There are the presence of applied elements and more ingenuity of graphic design.

冯嘉云

无锡上瑞元筑设计制作有限公司董事设计师/董事长
中国建筑学会室内设计分会高级室内建筑师
IFI 国际室内建筑师/设计师联盟会员
ICIAD 国际室内建筑师与设计师理事会会员
法国国立科学技术与管理学院项目管理硕士学位

Jiuli River Restaurant
九里河餐厅

设计单位：上瑞元筑设计制作有限公司
设　计　师：冯嘉云
参与设计：陆荣华
项目地点：无锡查桥九里河湿地公园
主要材料：大理石、电镀不锈钢、绿可木、皮革打印图案、墙纸打印图案、白影木、水曲柳浮雕板、老木头
建筑面积：2200 m²
竣工时间：2012.01

步入其间,积极阳光的空间印象顿生。形成这一印象,有对湿地公园大环境的符合,有"低投入、高品质"的甲方预期,更有设计师对"生态"空间的深刻理解与娴熟的手法展现。首先,在平面布局上,在完成餐饮的功能之外,突出了体验型、休闲型、国际感的现代业态气质;其次,材料使用上,环保型材料大量使用,既顺应低碳潮流,又与"湿地公园"保持了精神属性的一致;最后,通过丰富、明快的色彩运用、花卉等植物意向的造型,自然肌理的呈现,勾画出富于生机的就餐环境特质。

Step into the restaurant; customers will be impressed by the ecological, vital and active space which is with a sense of life. This impression accords with the over-all situation of wetland park, shows Party A's prospect of "Low Investment Brings High Quality Outcome" and demonstrates more the designer's profound understanding of "ecological" space and his skilled ways of demonstration. Firstly, in the aspect of layout arrangement, the design fulfills the restaurant's dining function and highlights its modern qualities of experiencing, leisure style and international style. Secondly, in the aspect of material use, the use of plenty of sustainable materials not only complies with the trend of low carbon, but also accords with the spiritual attributes of "Wet Park". And at last, plentiful and vivid colors and plant images such as flowers are used to present natural texture and delineate the features of vital dining environment.

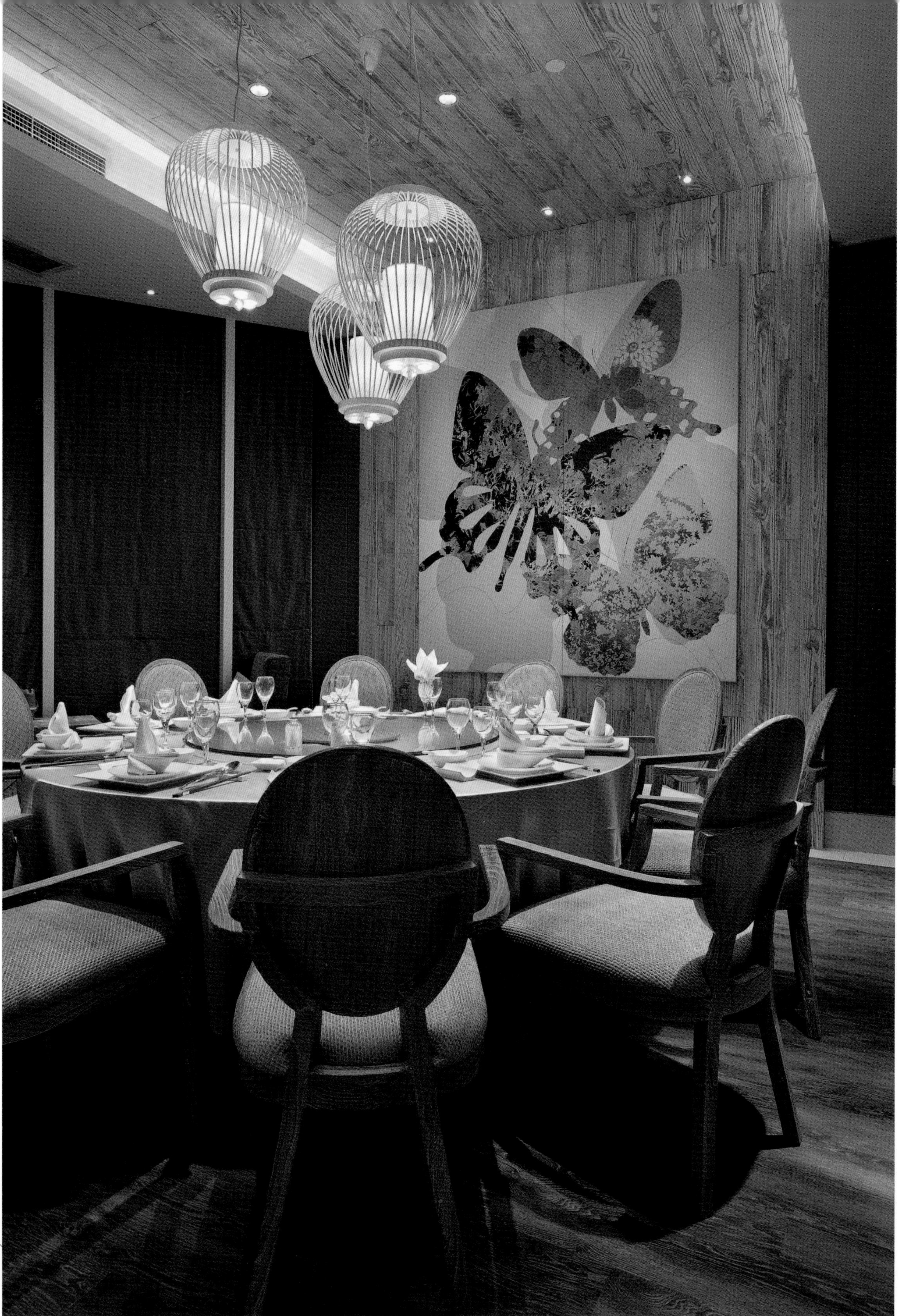

蔡宗志
台湾台南人
台北淡江大学建筑系
法国巴黎建筑专业学院
法惟思设计工作室(Atelier de L'A.V.I.)设计总监

温淼
内蒙古师范大学 环境艺术设计专业
现任 法惟思(北京)建筑设计咨询有限公司设计师
北京宫易锐思国际设计咨询有限公司设计师

Xisi Teppanyaki
西四铁板烧

设计单位:法惟思(北京)建筑设计咨询有限公司、北京宫易锐思国际设计咨询有限公司
设 计 师:温淼、蔡宗志
项目地点:西四羊肉胡同
建筑面积:1000 m²
主要材料:石材、玻璃、钢构、木材、硅藻土等
竣工时间:2012.03
摄 影 师:孙翔宇
撰　　写:蔡宗志

在建筑外观上，保留了这一时期的建筑材质，在胡同渐渐消失的现代社会中，能让各时期建筑保留且并存对整个都市的景观有种丰富性，更有其建筑历史发展的意义。处理的手法是不破坏的前提下在外面加上一层玻璃阳光房，使其建筑外观变成室内景观的一部分，还能欣赏户外的四合院。同时为了维持西四胡同区的整体形象，我们保留了门口的形式，做了一些现代手法的细部美化。

在室内风格上，自然粗犷简朴，造景是空间布局的另一个重点，希望让每个顾客在入内的行进间，欣赏不同景观角度，如每日一变的景观水池，大片粗犷的山石、山水纹路的洗石子墙。自然而未过分加工，虚实交叉堆砌出一种静谧的氛围，让顾客在用餐的同时也感受精神上的洗涤。

一楼包间是vip包间，力求简单舒适，每个包间都有主题墙，跟烧烤有关。二楼包间以能欣赏窗外的四合院胡同为主，偶尔能看到猫从屋脊上走过，窗外香椿树及松树错落分布在灰色四合院之间，白塔近在眼前。夹层的空间设计尽量不掩饰其原来的木结构，让空间更有个性。

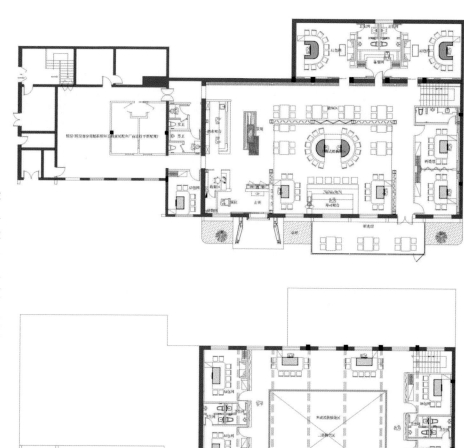

On architectural appearance, we still retain the architectural material quality during this period. In the contemporary society that lanes fade away, there will be a kind of richness in landscape and more significance on the historical development in architecture on the whole city to enable the architecture of all periods to retain and coexist. So, the solution is adding a glass of house outside without destroying, making the appearance on architecture be a part of landscape inside the room, further and decent appreciating the quadrangle dwellings outside houses. To preserve the whole image and in our insisting, the form of gate is maintained. Some contemporary measures on detail beautifying without destroying the globality of quadrangle dwellings will account for the times.

Taking natural and wild style indoors, spatial arrangement is another emphasis on landscaping. In the meantime of spatial experience, hoping that every customer could appreciate the different point of view during the intro-advancing. Like the changing landscape ponds every day, sheets of wild hanging rocks, flushing off the pebble wall with landscape lines, natural but not excessive processing, piling up the quiet atmosphere with true or false crossing ,and thus bring the spiritual washing-up to customers in the meantime of dinner as well as purify the moment of the busy urbanism.

The first floor is VIP rooms striving to be simple and comfortable, and each private room has subject walls related to barbecue. The second floor, focusing on appreciating the quadrangle dwellings and lanes, once in a while you can see a cat pass by on the ridge.Outside the window, Chinese toon trees and pines scattered distribute among the gray quadrangle dwellings, and the white tower is close at hand.

To enable the space to be more individual, prefer not to cover up the original wood structure on the layered space design.

佐佐木力
(SASAKI CHIKARA)

2001年3月 毕业于大阪工业大学工业部
2002年3月 infix（日本总公司）业务研修
2004年7月 上海英菲柯斯设计咨询有限公司
2007年4月 就任上海英菲柯斯设计咨询有限公司
　　　　　 北京事务所代表
2012年2月 就任上海泷屋装饰设计有限公司副总经理

Starry Night Dining

正村寿司HK-LokFu店

设计单位：Shanghai RID Co.,Ltd. / 上海泷屋
　　　　　 装饰设计有限公司
设 计 师：佐佐木力
项目地点：香港
建筑面积：130 m²
摄 影 师：Nacasa & Partners Inc.

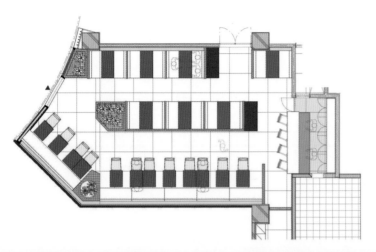

正村是位于香港购物中心内的家常寿司店。

作为日本料理的代表——寿司,在大多数人眼里还是一种门槛比较高的料理,而正村的经营理念就是做让大家在香港就能随意享受这一美食的店铺,营造"和式"的明亮、温馨氛围。

采用3色木材,将体现日本传统风格的格子,与现代风格相融合。格子,不仅能达到装饰效果,还能确保采光和通风,同时在一定程度上阻隔了来自外界的视线,传递传统古建的气息。极富规则性的格子在店内纵横交错、层层叠叠,营造空间整体的动感。而与墙面上的扇形土墙的对比,又突出了空间的进深感。

木材的温和,能将人的感官柔化。我们最大的成就感,来源于能让顾客与亲朋、好友、自己的心爱的人在这里一起度过欢快、愉悦的时光。

Located at shopping center in HK, Zhengcun is a homely sushi store.

As a representative of Japanese food-sushi, within the most people's vision, nevertheless, it has a relative higher threshold, while the management idea of Zhengcun is that it will exactly enable every one to enjoy the good food at will-a good food store, to build the atmosphere both of bright and warm of "He" style.

Using wood which having 3 colors, which will embody traditional Japanese style lattices, mixing together with modern style. Lattices, it not only can achieve decoration effect but also insure both lighting and venting, meanwhile, having obstructed the sight outside to some extent, delivering flavor of traditional ancient architecture. Highly regular lattices cross with multiple layers to build integral innervations of spaces. While contrast to sector cob wall of wall space, it will highlight the feeling both of entering and being deep as well.

Temperate wood will soften sensory organs of humans. It will be the maximum sense of achievement for us to enable customers to spend time both of lively and joyful with relatives, friends, darlings of its own together.

周易

1959年 出生于台中市
1979年 以自学方式学习建筑与室内设计
1989年 创设JOY室内设计工作室
1995年 创设JOY概念建筑工作室

Karuizawa on Gongyi Road

轻井泽公益店

设计单位：周易设计工作室
设 计 师：周易
参与设计：陈威辰
项目地点：台湾台中市
建筑面积：1117.3 m²
主要材料：铁件、铝格栅、文化石、铁刀木、南非花梨木、玻璃
竣工时间：2012.01
摄 影 师：和风摄影　吕国企

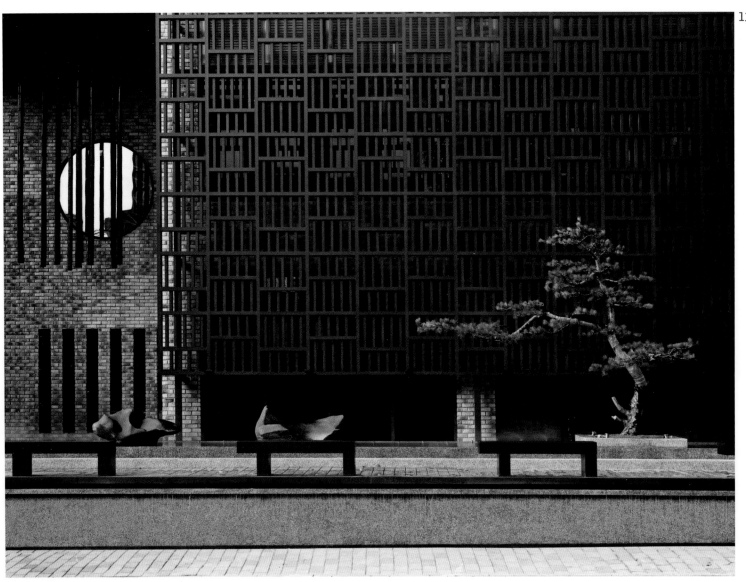

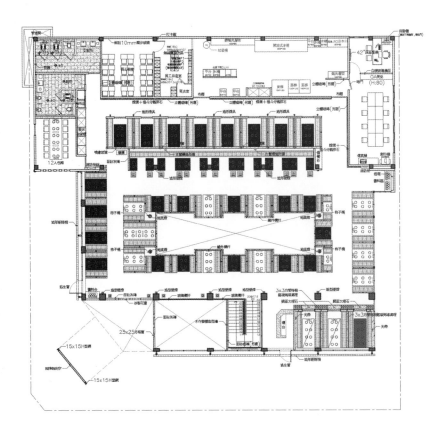

建筑要有意义,不仅在乎工艺内容;更仰赖整体文明、历史、气质的传承。本案延续古朴宏伟的建筑特色,除了静谧禅韵;更多了一份源自悠远中国的人文深度。

基地坐落两路交会的角地,锐角斜切后成为六角形入口与主要店招的展示面,在基地因应地势略行垫高的基座上,超过7 m高的三面黑灰色建筑外观非常有特色,首先是大面铸铁精工打造的倒L型店招+雨遮,店招正面嵌上书法名家挥毫的巨大白色"轻井泽"铁壳字,夜间在灯光衬托下视觉张力格外鲜明。

沿着架高基座外缘为兼具等待区机能的木栈景观步道,步道与建物之间规划镜面水景,点缀嶙峋的巴东石、烛台灯和蒸腾水雾,并贴心设置别致长凳可供来客小憩。在架高基座外缘与马路的落差处,结合循环水幕与灯光,打造流动而梦幻的光瀑意象,夜幕时分,整座建物仿佛漂浮在光托之上。

静定稳重的建筑主体采用古色古香的斑驳灰砖砌作,两侧立面分别装置大小交错衔接的黑色格栅以及纤长的直列开窗,在宛如古代仓库或碉堡建筑的恒久时间感中,

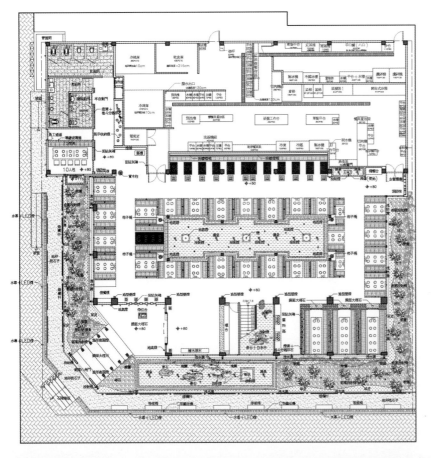

融入中式窗花简化后的格栅线条，低调禅风里，同时摇曳着绿竹的写意以及五叶松的迎宾热情，让这出落在现代的大器建筑，经由设计者绘景叙情的技巧，让画面回荡于飘渺水墨中国与和风浸润之间，反复搓揉出越醇越香的人文风味。

　　入口大门延续外廊的六角门拱，淡雅的苏州庭园意象隐约生成，内部分别为4 m、3 m的二层楼面，分别规划包厢与散座，总计400余的座位，却能避免人声杂踏相互干扰，维持私密又有情调的餐叙氛围。内部空间设定延续大量实木格栅与情境光源，视觉上融入处处可见的古玩书柜墙、大红色喜气洋洋的鸟笼灯饰、形成空间天井并串连上下两个楼层的水景设计，架构三度空间处处有景的绮丽逸界，多处墙面装置运用渐层玻璃与喷绘手法，透过晕染酝酿独特的"光迷幻"效果，柔软而忠实地传达水墨书法与精致图章代言的人文书香，整体层次万千的情境铺陈与灯光布局，透过多种自然材质与高难度工法，凝聚一种跨越时间长廊的精湛演出；当然也让五感在内的享受更有深度。

Meaning of constructions does not lie only on technologies, but more on the inheriting of comprehensive civilization, history and qualities. This project continues the architectural features of primitive simplicity, majesties and also cultural profundity originating from age-old China, besides serenity and Buddhist's charm.

The restaurant which is located at the joint place of 2 roads is featured by hexagonal entrance formed by chamfering an acute angle & displaying surface of main signboard of the restaurant. A building of over 7m with three black grey facades is located at the foundation support raised because of topography and the architectural fa?ade of the building is with great features. The first features are the reverse L-shape signboard made elaborately by cast iron and the rain awning. The front surface of the signboard is embedded with large, white and ironclad word "Karuizawa" written by calligraphy master. Being foiled by lights at night, the word is with especially bright vision effort.

There is a wooden landscape footpath with a role of waiting area at the outer edge of heightened foundation support. Planned water landscape between footpath and the building is adorned with jagged Padang stones, candlestick lamps and rising mist and unique benches are elaborately equipped for guests to rest. Zhou Yi, the responsible designer of the project, once said: light is the soul of constructions. The sentence points out frankly the reality that it is light & shadow that determine the true essence of the circumstance. The most special of the project is the difference of level between the outer edge of heightened foundation support and the road. It integrates circulating water screen and light and creates flowing dreamlike images of light waterfall. When darkness gathers, it seems that the whole building is floating above the light.

Peaceful and sedate principal part of the construction is made of old-timey mottled grey bricks and facades at both sides are equipped with black intersect-

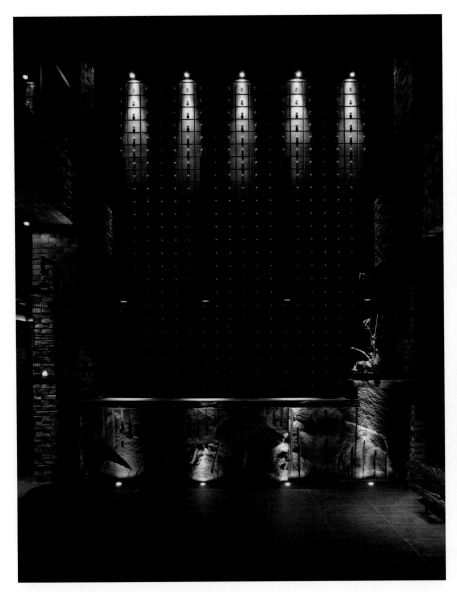

ing grilling and long orthostichous windows respectively. It is just like integrating grilling lines simplified from Chinese style paper-cut for window decoration into the lasting sense of time of ancient warehouses or blockhouses or is just like the enjoyableness of green swaying bamboo and white pine's enthusiasm of guest welcoming in silent gentle wind. This enables the modern outstanding construction to express repeatedly the more and more mellow cultural aroma in the immersion of wash painting and gentle breeze by the designer's techniques of scenic painting and passion expressing.

The entrance door designed with hexagonal gate arch as that of outer outline faintly generates quietly elegant image of Suzhou gardens. The interior space is designed into two floors of four-meter-height and three-meter-height respectively. The space is designed into compartments and odd seats which provide over 400 seats in all and the design can not only help avoid mutual interference of voices and noises, but also maintain private and emotional dining and chatting atmosphere. The interior space design uses continuously plenty of solid wood grilling and circumstantial light sources. In vision, the building blends in antique and bookcase walls everywhere & red jubilant decorative birdcage lighting, forms space courtyard and water scenery that connects the upper and lower floors, creates gorgeous leisure three-dimensional space that has views everywhere, applies gradient glass and spraying methods on many wall devices, creates unique "Light Illusion" effect through color infiltration to express gently and faithfully cultural elegance represented by ink painting handwriting and exquisite seals. By various natural materials and highly difficult solutions, the whole circumstance elaborating tens of thousands of layers and light layout creates an exquisite performance covering different periods and of course, brings customers more profound and impressed sense enjoyment.

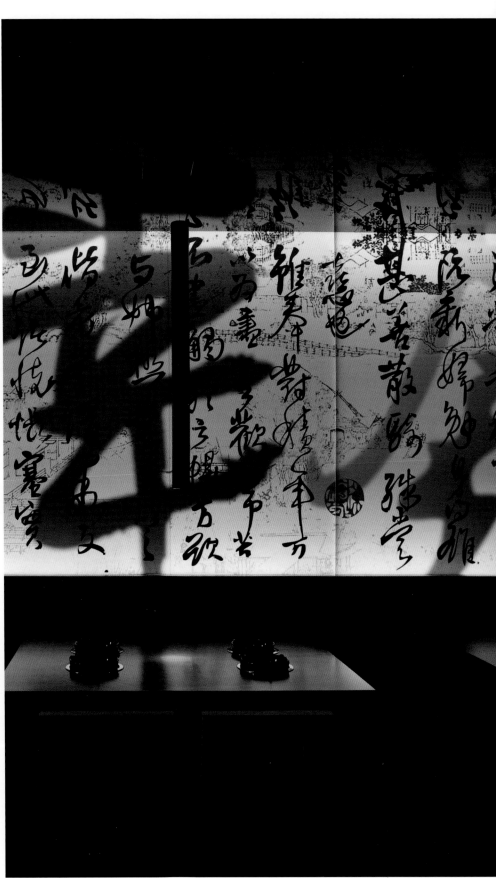

于丹鸿

重庆朗图室内设计有限公司设计总监

朴素餐厅

Simplicity restaurant

设 计 师：于丹鸿
参与设计：李响
项目地点：重庆市江北区北滨路龙湖星悦荟6-5F
建筑面积：500 m²
摄 影 师：李响

朴素餐厅是一间素食餐厅,提供绿色自然的纯粹素食餐饮,朴素者,天下之大美。这个设计中,试图去营造一个可以让身心片刻宁静,抛开当下的喧扰,可以安静地去寻找每个人本真需求的地方,人和自然达成更和谐的关系。通过设计去讲述一个朴素的生活哲学。

设计意图用尽可能少的设计语言,去表达含蓄内敛的东方文化。期望通过没有具象信息诉求的方式,更多的是暗示和含蓄地表达。收和放,是空间布局的重点。通过起伏蜿蜒的隧道,到豁然开朗的穿顶。从临窗见江的开放明亮,到竹栅围合的私密幽暗。通过收放的布局去引导心理感受的变化。

设计中仅用到两种主要材质,竹和石。江西的细竹片条,制作了围合隔断,隧道,穿顶,地面也是铺贴的同样的竹条。福建的灰色花岗岩石,保留了刚刚开采出来时的劈离石面,无修饰的上墙,通过自然的起伏展现丰富的效果。

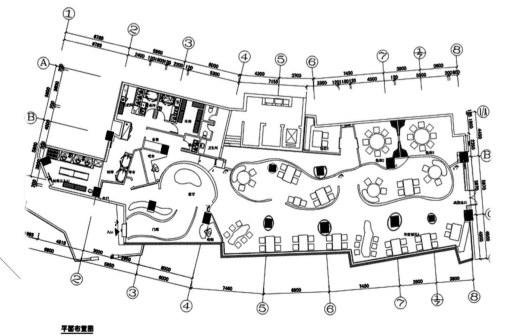

平面布置图

Simplicity restaurant is a vegetarian restaurant, providing pure vegetarian catering both of green and natural, one who is simplicity, which will make world beautiful. The design tries to build a place which would make both the heart and the body be moment's peace, and putting aside immediately pother, which would make one peacefully seek for genuine requirements of oneself, humans would make harmony relations with the nature, telling a life philosophy of simplicity through design.

The purpose of the design would use design language as little as possible to express eastern culture of reserved style. Hope to reservedly express more hints without appeal of concrete information. Both putting away and release is the key to the layout of the space. From winding tunnels which are both ups and downs to domes which make one suddenly enlightened, from both brightness and openness which is nearby a window to both privacy and gloomy which is surrounded by bamboos, to guide the change of mental feelings through the layout both of putting away and release.

There are mainly only 2 materials such as bamboo and stones during design. Fine bamboo chip from Jiangxi had made obstructions which are used to surround, tunnels and domes, and ground is also paved with the same bamboo chip. Ashy granites from Fujian had maintained split stone surfaces, wall run without modifications, which had showed abundant effectiveness through both ups and downs of spontaneous.

潘鸿彬

泛纳设计事务所创始人
香港理工大学设计学院助理教授
香港室内设计协会副会长
IFI国际室内建筑/设计师联盟执委

Starry Night Dining
星光捌号

设计单位：PANORAMA 香港泛纳设计师事务所
设 计 师：潘鸿彬、谢健生、蔡志娟
项目地点：无锡
建筑面积：700 m²
主要材料：紫红色光纤、枕木水泥壁、生铁、
　　　　　黑白牛图案
竣工时间：2012
摄 影 师：吴潇峰

星光捌号是无锡市中心新开设的一间牛排餐厅,位于历史活化项目"西水东"的百年工厂房之内。设计概念是将富有历史价值的仓库加以改造更新,保留其大部分建筑的原有风貌。加建的夹层结构大大增加了可用空间,使它成为时尚浪漫的现代化餐饮地标。设计策略是将传统的牛排屋餐饮体验提升到一种新的境界。餐厅的两个用餐区内多种类型的座位布局迎合不同顾客的要求,在这时尚而浪漫的意境中享用锯扒之乐。

　　中庭用餐区保留原有旧建筑的元素,包括10 m高屋顶的斜面天窗和梁柱结构、外墙红砖和室内水泥壁及窗框等,使顾客可以一边进食一边欣赏老建筑的新面貌。2 m高的雄牛屏用刀叉组成,是餐厅的品牌标志,使顾客一见顿生食欲,区内垂直宽敞的空间里,6 m高镜钢铁酒柜和VIP区顶上的水晶吊灯是装饰重点,中央和周围用餐区着真皮梳化,天花垂吊着紫红色管线造成的点点星光,缀成了星光下的晚宴。

　　夹层用餐区新建的钢构夹层是半开放式。其多功能区的黑白牛图案吊顶和地灯使顾客到此有宾至如归之感。半透明的活动屏风装置提供活动的间隔,以便满足举行不同的活动需要。最后顾客在黑镜饰面的洗手间里结束了其星光之旅,留下了难以忘怀的印象。

StarEight is a new steak house situated at one of the warehouse buildings within a 100-year old factory compound in the city of WuXi, China. A historical re-vitalization design strategy was adopted to preserve most of the original building features with addition of new mezzanine structure to maximise the spatial potentials and turned the old warehouse into a hip F&B landmark. A main scene of "Starry Night Dining" was created to move the traditional Steak House dining experience to a new level of trendiness and romance. Various seating patterns were created in the two dining zones to cater for different customer needs.

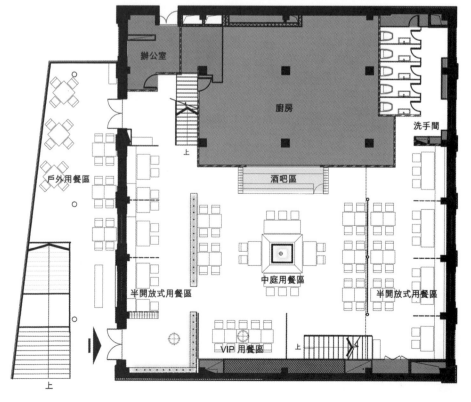

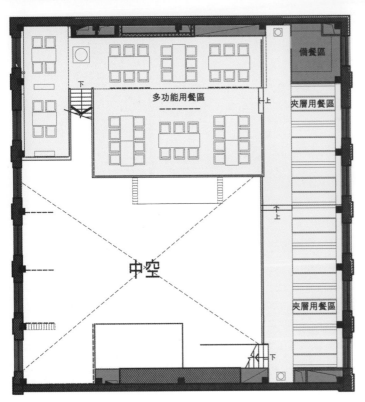

Atrium Dining Zone – Original building features of the 10m high pitched roof skylight, truss structure, exterior brick & interior fare-faced cement walls, window frames were well-maintained and exposed to the eyes of the customers.

2m high bull-shaped partition made by knifes & forks gave clue and defined brand identity to the restaurant. Verticality & spaciousness of the atrium dining zone were emphasized through full-height mirror stainless steel wine rack and crystal chandelier above the VIP table. Ceiling suspended fibre optics in violet colour created starry night effect to the leather-upholstered central & periphery booth seating areas.

Mezzanine Dining Zone - New built steel structure mezzanine floor was introduced to provide a semi-enclosed dining zone. VIP area with black & white bull-patterned ceiling-wall and floor lamps gave cosy & home feelings to the space, sliding semi-transparent curtains provided flexible partitioning for holding different events. The romantic starry night experience was finally completed at the black-mirrored washrooms and left an unforgettable memory to the customers.

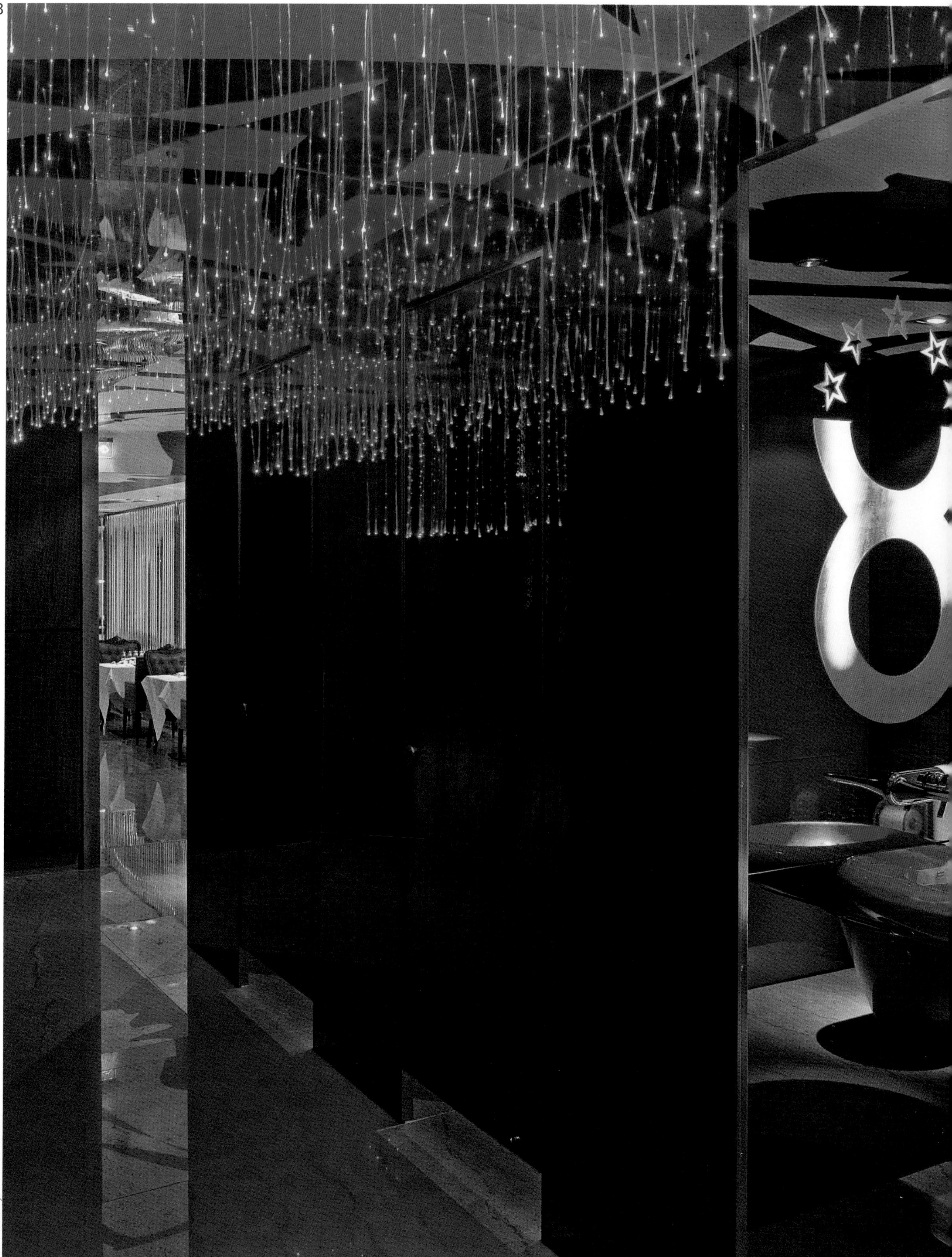

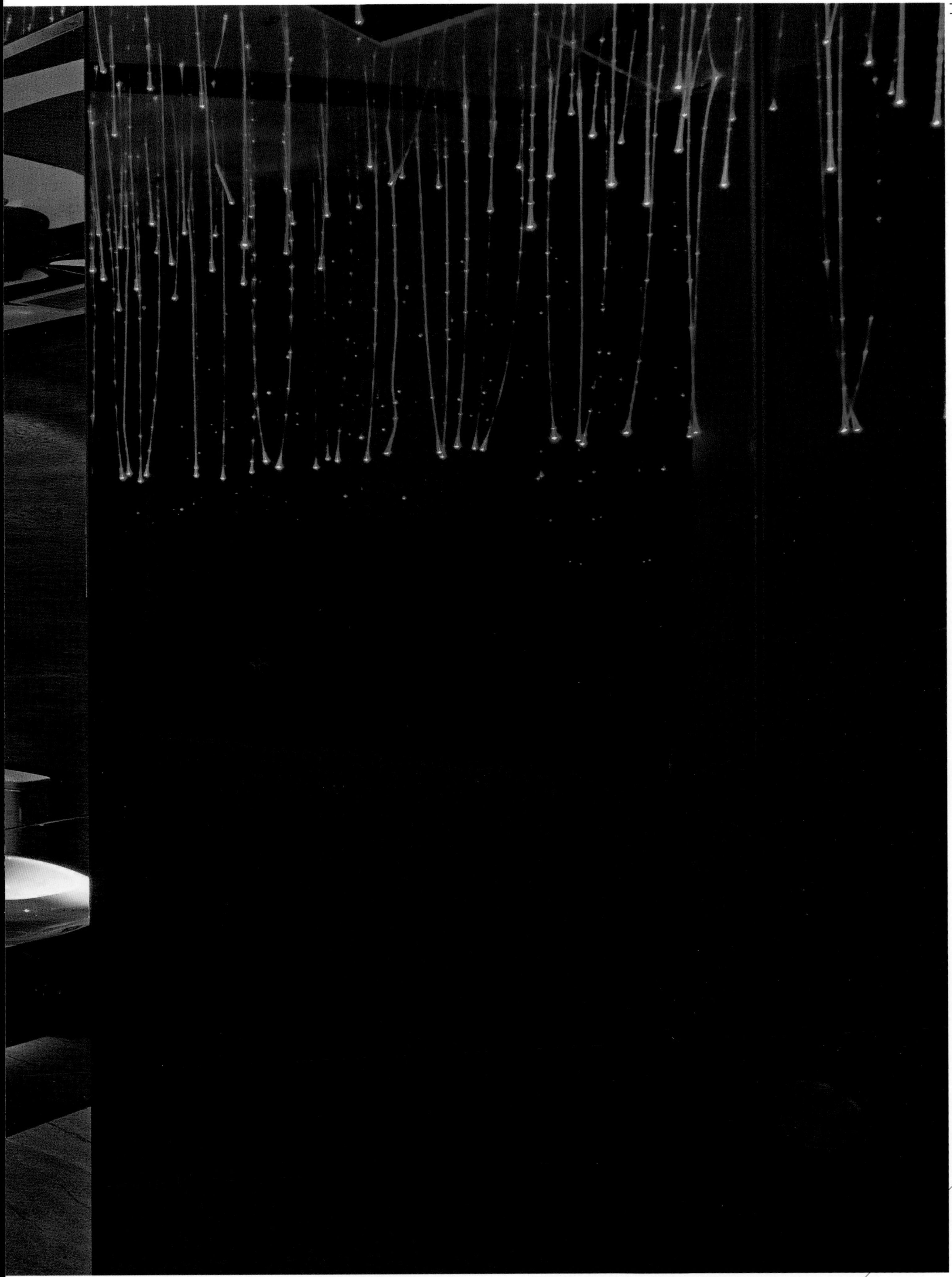

徐晓华

中国建筑学会室内设计分会会员,国家注册室内建筑师。

毕业于苏州工艺美院,苏州徐晓华室内设计有限公司设计总监,擅长于酒店、餐饮、娱乐、休闲等大型娱乐场所的室内设计。

Yangzhou Fulin Hu Garden
扬州富临壶园府邸

设计单位:苏州徐晓华室内设计有限公司
设　计　师:徐晓华
参与设计:苗永光、李爱岭
项目地点:扬州广陵区
建筑面积:1800 m²
主要材料:石材、地板、成品木饰面板、
　　　　　布艺硬包、墙纸
竣工时间:2011.08
摄　影　师:潘宇峰

壶园是一座极具江南特色的私家园林,始建于清代,设计师在其室内设计构思时也结合这一特色,加上现代生活学,给人风雅、舒适的感觉。餐厅地面是深浅纹石材辅以实木拼花地板,深色的圆型柱子加上订制的艺术灯具等复古元素的点缀,整个空间流露出高雅与低调的奢华感。包厢里摆放造型简单、颜色淡雅的家具,复古的陶罐摆设,古典韵味的玉器、贝壳等软装的陈设使一种宁静、高贵、雅致的感觉在整个空间中蔓延。庭院、走廊设计师运用新中式的表现手法,让整个空间又多了一份优雅的文化气息,置身其中,让人流连忘返。

Hu Garden is a most Jiangnan private park. It was built in Qing Dynasty. The designer has merged this character into the space. Coupled with modern life, it gives people a sense of elegant and comfortable. The restaurant ground is made of stone and solid wood parquet floor. The dark round pillars coupled with the customed art lamps and other vintage decorative elements, the entire space seems so elegant and luxury. In the room, there is placed simple elegant furniture, retro pottery and classical jade, shells and so on. They are showing a kind of peace and noble sense, which also extend to the whole space. For the courtyards, corridors, the designer used new Chinese technique to express, letting the space add an elegant cultural atmosphere, making the space full of charming.

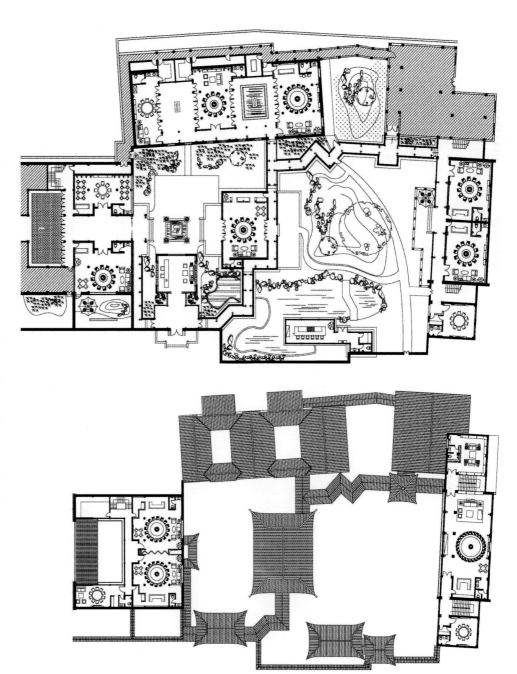

HASSELL

HASSELL 是一家国际化的专业设计公司,拥有14间事务所,分布于澳大利亚、中国、东南亚和英国。公司有超过900名员工,业务经营已有70年历史,涉及全球多类市场。

HASSELL 作为一家跨专业设计公司,拥有多领域专业实力,包括建筑设计、室内设计、景观设计和规划设计,公司注重可持续发展理念与设计实践的结合,兼备卓越的城市设计能力。

作为一家私营合伙公司,HASSELL 的每一家事务所都具备了充分的灵活性和自主性,可同时为本地客户和国际客户提供服务,公司的综合资源和集体专业经验则为各地工作室提供了无可比拟的优势。

2012年世界建筑设计(BD World Architecture)公布的世界建筑设计公司100强中,HASSELL 在澳大利亚设计事务所中排名第一位。HASSELL 已获得超过650项设计大奖。

Assaggio Trattoria 意大利餐厅

设计单位:HASSELL
项目地点:中国香港
建筑面积:600 m²
设计时间:2011
摄影师:Graham Uden

Assaggio餐厅位于香港的艺术枢纽——香港艺术中心6楼，面向开阔区域，维多利亚港全景在这里一览无遗。当顾客走进Assaggio Trattoria意大利餐厅，即被入口处的比萨与意面吧台所吸引。醒目的艺术作品置于乡村风格石材饰面之上，木纹深色调营造了沉着且愉悦的用餐环境，让人联想起舒适而友好的意大利食品专卖店。意大利文Assaggio解作"品尝"或"一口"的意思。设计概念意在打造一个"用餐舞台"，充分展现一种意大利杂货店式的舒适惬意氛围。粗糙的烧面砖和深色木地板与色彩鲜明的艺术品让食客暂时摆脱繁忙的生活并进入一个充满艺术氛围的就餐环境。

Being located on the 6th Floor, Hong Kong Arts Centre—art hub of Hong Kong, Assaggio Trattoria Italiana faces open areas and can have full view of Victoria Harbor and gets everything of the harbor at a glance. As soon as customers step into Assaggio Trattoria Italiana, they will be attracted by the pizza & pasta bar counter at the entrance. Striking art works are places on decorated stone surfaces of village style and the dark wood grain creates equanimous and pleasant dining environment and lets people connect comfortable and friendly franchised Italian food stores in mind. The Italian word Assaggio means "taste" or "a bite of". Design conception of the restaurant aims at constructing a "dining platform" and demonstrates fully comfortable and satisfied atmosphere of Italian grocery style. Rough bricks, dark timber floors and colorful works of art help eaters get rid of busy life temporarily and get into a dining environment full of artistic atmosphere. The restaurant enjoys great popularity among customers and eaters since its starting business in March 2011.

Concrete

荷兰著名设计公司,Concrete 的整个团队约有35位职业人士:他们是视觉市场商人、室内设计师、平面设计师和跨学科团队中参与项目建设的建筑师。

Hotel Castell D'emporda
castell d'emporda 餐厅

设计单位:Concrete
设 计 师:Rob Wagemans, Ulrike Lehner, Marc Brummelhuis, Sofie Ruytenberg, Femke Zumbrink, Erik Van Dillen
项目地点:西班牙
主要材料:雨伞、白漆刷钢珠、透明窗帘、大理石桌、皮革沙发
图片提供:concrete architectural associates, Ewout Huibers, Robert Aarts

位于西班牙吉罗纳的格泰吉达酒店有一个大露台,在这里可以饱览胜景。露台之上的特色餐厅,客户希望建筑师设计一个封闭的,遮蔽风雨的,但又仅仅像是飘在露台之上,并且与历史建筑和谐的建筑。

　　一般来说,在室外阳台空间享受自然,人们需要一个遮挡风雨的大树,这样流线和视线几乎不受影响。从这个考虑出发,建筑师建立12个雨伞随机放置在阳台上,相互交叠,外层是透明轻质可移动表皮。树立起梧桐树般的形象,轻盈漂浮,与古建筑融合。如果将建筑处理成玻璃屋顶,就像是温室那样,这将与古建筑的融合感大打折扣。阳伞的顶部与侧边做出生锈的效果,与古建筑很搭调,下面是白漆刷钢柱,空间被营造出开放明亮的户外空间感。建筑可以保持开放,但是在寒冷和起风时,也能在1、2分钟之内快速的闭合漂亮透明的窗帘。里面摆放着精心设计的圆形和方形大理石桌,还有白色皮革休息沙发。这个在大遮阳伞下的远行户外酒吧叫玛格丽塔。

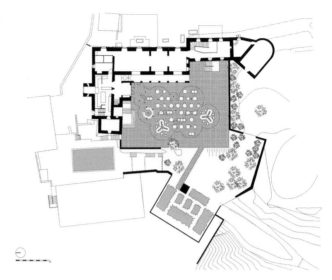

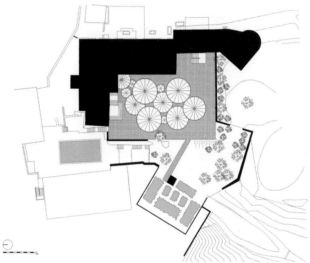

Hotel Castell D'emporda located in Girona, Spain offers a signature restaurant including a large terrace with great views over the surrounding landscape. Concrete designed, at the clients' request, a roof or covering for this terrace with the possibility to create an enclosed space with full wind and rain protection. One of the design conditions was to create a covering that works in harmony with the historical and listed building. Additionally we wanted to maintain the terrace feeling while be seated under the covering.

In principle a terrace is an outdoor space where one can enjoy the weather. If necessary, you need a parasol for sun or rain protection, but there is almost no obstruction between the visitor and the view. The solution was to create abstract parasols. 12 Circles in divers diameters are placed randomly on the terrace. Where the circles touch they melt together, the open spaces between circles are filled in with glass. The circular parasol shapes enhance the feeling of being in an outdoor environment on a terrace. The shape of the covering appears as a separate almost temporary element, leaving the ancient building untouched. A glass roof or a winter garden would to much become a building, create a feeling being inside a structure and would also appear as an extension of the building, damaging the ancient character. The top and edge of the parasols are made in rusted steel, seeking harmony with the ancient building and the natural environment. The white painted steel columns and ceiling create an open and light outdoor atmosphere under the parasols. Transparent sliding curtains can be hung easily in colder periods but always stay open. When the mistral winds suddenly appear the whole terrace can be closed in a couple of minutes. Round and square marble tables and two white leather lounge couches create different seating facilities. Underneath one parasol a circular outdoor bar is placed. The restaurant now has his own name: Margarit.

蒋建宇

宁波宁海人
2001年组建大相艺术设计公司
2011年组建大相莲花陈设艺术公司

Beijing Financial Street Continental Hotel Xin Rong Ji Restaurant
北京金融街洲际酒店新荣记餐厅

设计单位：杭州大相艺术设计有限公司
设 计 师：蒋建宇、郑小华、胡金俊、李水、楼婷婷
项目地点：北京市西城区金融街11号
建筑面积：2700 m²
主要材料：京砖、银龙灰大理石、珍珠黑花岗岩、柚木、硅藻泥、铜片
设计时间：2011
摄 影 师：贾方

餐厅位于北京金融街洲际酒店内,地段繁华而高贵。但本餐厅却给人以"自然、空灵、沉静、朴素"之感受,在空间的意境上力求幽玄空灵的精神之美。

在材质的应用上,选择了天然的材料,而且尽可能保留自然材质的天然纹理和质感。比如原木、土砖、竹子、石板、溪石、藤席。利用温润的材料特性营造朴素、内敛的气息,调和人与物、人与空间的和谐。

在空间表达上力求简单、素美,崇尚朴实、自然、亲切,去除过多的无关装饰元素。尽可能减少设计痕迹,使消费者从自然朴素的形态中体验一种幽玄之美,使就餐的空间亦成为人们躲避烦杂世俗的栖息之所。

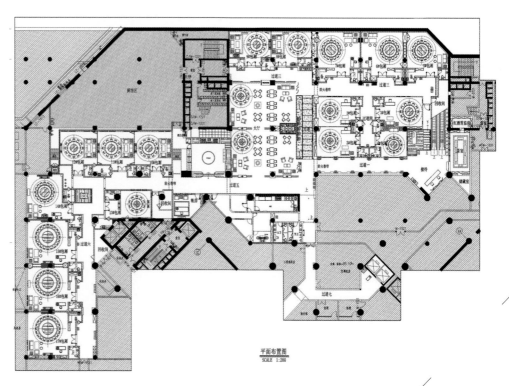

平面布置图
SCALE 1:200

The restaurant is located in the Beijing Financial Street Intercontinental Hotel, where is noble downtown area. While, the restaurant brings people a natural, ethereal, silent, and pure feeling, as well as, unseen ethereal beauty of the spirit in the space of artistic conception.

In order to retain the natural texture as much as possible, the materials of the restaurant are entirely natural, such as, wood, bricks, bamboo, stones, and rattan mats.

The restaurant makes full use of mildness of materials to create restrained atmosphere and harmony among people, space, and objects.

In the spatial expression, the restaurant tries to show its simplicity, beauty, and gentleness without too many irrelevant decorative elements and to reduce the design traces as much as possible, therefore, making the consumption a journey to experience the unseen beauty from the natural and simple form, as well as, making the dining space a secular shelter for consumers to avoid troubles.

王奕文

和合堂设计品牌创始人\设计总监

蓝犀装饰公司设计总监

Spring feast

春江宴

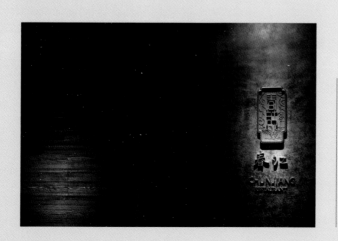

设计单位：和合堂设计咨询
设 计 师：王奕文
参与设计：胡岩峰
项目地点：北京中粮广场
建筑面积：900 m²
主要材料：石材马赛克、地板、壁纸、皮革、烤漆、
　　　　　镜框、绘画作品
摄 影 师：郑尧琛

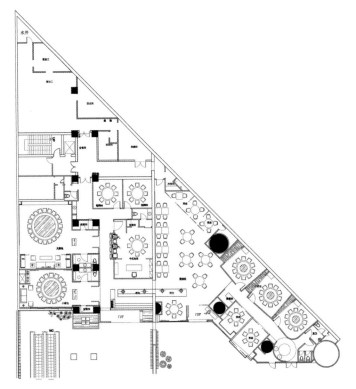

位于北京中粮广场的春江宴,以《春江花月夜》的氛围,营造出优雅,内敛,低调奢华的时尚餐厅。

　　五个大包间为高端客户量身定做,"江天一色无纤尘,皎皎空中孤月轮",虽无玉雕粉饰,但看似简单平实的灰色,却因为丰富的层次,特别的图案,严谨的配色,震撼的绘画作品,将"低调的奢华"演绎得淋漓尽致。人虽未至诗中景色,然心已向往之。

　　"鸿雁长飞光不度,鱼龙潜跃水成纹",大厅的鱼灯赋予这内敛的空间些许期盼之意。绿色的座椅,绿色的纱幔,绿色的荷花,暖色的镜框,仿佛镶嵌着一个月光下温暖的画面迎接远行的游子归来。

　　三个实用性很强的包间为朋友小聚提供了一个安静的所在,当隔断徐徐打开时,那排列整齐的出自艺术家之手的54幅画面,空间尽端的流水坛,轻纱环绕的三个空间印入眼帘。好一副"江流宛转绕芳甸,月照花林皆似霰"的美好意境。

"春、江、花、月、夜",赋予了这个含蕴、隽永的空间灵动的,深邃的艺术感受。

Located in the Beijing COFCO Plaza, the spring feast makes full use of the atmosphere from the song A Moonlit Night on the Spring River to create an elegant, restrained, and low-key luxury fashion restaurant.

Five large dining rooms are designed for high-end customers, "a sky color river does not have dust, while the moon is bright and silent", although without luxury jade decoration, the five rooms with seemingly plain grey are equipped with rich color level, special designs, precise color matching, and shocking pictures to interpret the low-key luxury to its best. Customers who come here couldn't help yearning although haven't really arrived at the poetry scene.

"Wild goose flies light, diving fish water lines" The fishing lamp of the hall gives the reserved space a bit of hope. The green chair, green veil, green lotus, and warm color frames are just like forming a warm moonlight picture to welcome the arrival of travelling man.

Three quite practical rooms has provided a quiet place for friends reunion. When the cut opens gradually, 54 pictures from artists come into your sight neatly, followed by space water altar and three gauze-wrapping spaces. How nice the atmosphere it is!

"Spring, river, flower, moon, and night" has given the full and meaningful space with smart and deep feeling of art.

利旭恒
出生于中国台湾
英国伦敦艺术大学荣誉学士
古鲁奇公司设计总监

Charme Restaurant A Dream of Dragon at Hongkou, Shanghai

港丽餐厅上海虹口龙之梦店

设计单位：古鲁奇建筑咨询(北京)有限公司
设 计 师：利旭恒
参与设计：赵爽
项目地点：上海虹口龙之梦
建筑面积：650 m²
竣工时间：2012
摄 影 师：孙翔宇

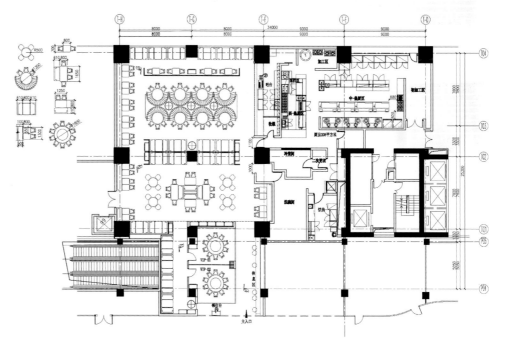

来自香港的港丽餐厅是一家专营港式料理的品牌，在北京与上海都已有大量的粉丝，来自台湾的设计师利旭恒给予了本项目一个有趣的概念："未来世界"。这也是一部40年前的电影。

电影情节进入了空间，机器人布下天罗地网的搜抓人类，人们如同马戏团表演般的在舞台上逃窜，当人类不幸被捕抓之后机器人利用输送带将人类送往另一个世界。

设计师延续未来世界电影概念主题，影射剧情的网状物，输送带，垂直装饰物转化在空间中。从天花一颗一颗的镜面玻璃球，延伸的以多层次灰白基调的墙面造型壁面，对比之后以墨黑皮制座椅作为视觉所及的句点。空间的中央区，吧台造型以未来世界的弧形语汇优雅地呈现科技美学，镜面底板搭配白色人造石桌面构成吧台，简洁的材料语汇，呈现一种未来的时髦美感，吧台的对面是一个大玻璃盒子，内部为条通往地下层的输送带型电扶梯，在餐厅里可以透过玻璃看见双向流动的人群，呈现了未来世界电影情节中的黑色幽默。

设计师为餐厅墙面所制作的装置艺术以"垂直流动"为概念主轴，利旭恒以多层次白灰基调的人造皮革管子，构成一连串的垂直视觉体验，这件雕塑概念的墙面装饰结合了装置艺术的概念，由餐厅入口到用餐区，成为餐厅室内空间的皮肤。

港丽餐厅成功的将电影情节，艺术，设计融入在同一个空间中，成为一体。

Charme Restaurant which is from Hong Kong is a brand that specializes in providing Hong Kong cuisine and it has won plenty of fans in Beijing and Shanghai. Taiwanese Designer Lee Hsuheng gives the project an interesting concept "Future World". The concept is also name of a film produced 40 years ago.

Plot of the film gets into the space and robots spread dragnet to search and grasp human who are like circus performers fleeing in disorder on the stage. When human are arrested miserably, they will be sent to another world by conveyer belts.

The designer continues the concept of Future World and mesh, conveyor belt and vertical ornaments alluding to the plot are used in the space. One and another mirror glass balls on the ceiling and extended multi-level wall of gray tone lead to the full stop of visual sense, the black leather-made seats after comparison. The central area in the space, modeling of bar counter and the arc vocabulary of the future world elegantly demonstrate the beauty of science and technology, while the bar counter consisting of

mirror face baseboard and while artificial stone table and concise materials present fashionable beauty of the future. And the large glass box, which is opposite to the bar counter and whose interior is a conveyor-type escalator and from which bidirectional flows of crowds may be seen, presents the black humor expressed in the plot of Future World.

Art installation designed for walls of the restaurant by the designer regards "Vertical Flow" as the principal concept axis. Lee Hsuheng uses multi-layer artificial leather of pale grey keynote to create a series of vertical visual experience. This wall decoration with sculpture concept integrates with concept of art installation and forms skin of the interior space of the restaurant' from the entrance of the restaurant to the dining area.

Charme Restaurant successfully integrates plot of the film, art and design into one and into the same space.

王砚晨(Wang Yanchen)
毕业于中国西安美术学院
意大利米兰理工大学国际室内设计硕士
经典国际设计机构(亚洲)有限公司 首席设计总监
北京至尚经典装饰设计有限公司 首席设计总监
中国建筑学会室内设计分会 会员

李向宁(Li Xiangning)
意大利米兰理工大学国际室内设计硕士
经典国际设计机构(亚洲)有限公司 艺术指导
北京至尚经典装饰设计有限公司 艺术指导
中国建筑学会室内设计分会 会员

Wong's Chafing Dish Restaurant at Yizhuang
王家渡火锅亦庄店

设计单位：经典国际设计机构(亚洲)有限公司
设 计 师：李向宁、王砚晨
项目地点：北京
建筑面积：1700 m²
主要材料：夜里雪、麻砂、做旧实木地板、实竹贴皮、印刷玻璃
竣工时间：2012.07

坐落于北京亦庄的王家渡火锅强调人与自然的和谐。设计师希望走近餐厅的人们能够去感受自然、发现自然，这都归功于现代空间中对传统元素的当代运用。

整个项目的进程既是一次对空间的整合又像是进行着一场再造自然的活动。连续不断延伸的白色石材墙体之上，镌刻的鱼儿带给人们自由与轻盈，白色的天花像微风吹拂的水面，泛起层层波浪，通道两侧大气的通体玻璃表层上，数字化的水墨荷花淋漓酣畅。素白的门板荡起一圈圈的涟漪，而形态各异的鹅卵石，通过不同的材料穿插于空间之中。这诸多的自然意象在室内各个空间之间相互渗透、彼此呼应，通过再造自然空间的手法使环境更加巧妙地融合在一起。

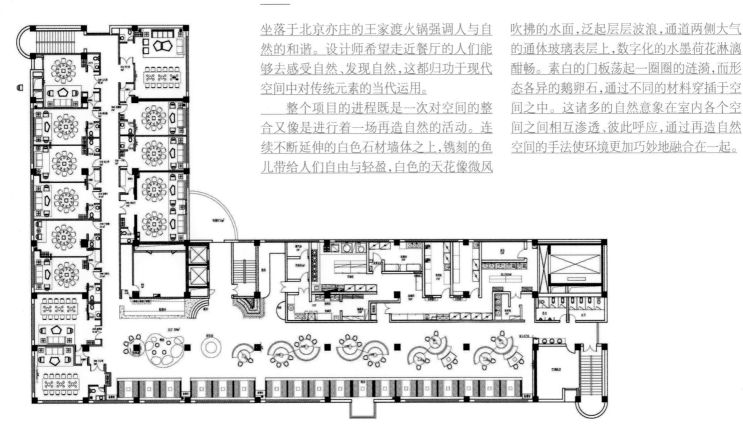

三层火锅平面图

All originates from the respect to the beauty of the nature that requires great unification of sociality and naturality. Wong's Chafing Dish Restaurant which is located at Yizhuang, Beijing emphasizes on the harmony of human beings and the nature. The designer hopes that people who come near the restaurant can feel the nature and discover the nature with their own eyes. And all these are achieved by the contemporary use of traditional elements in modern space.

Process of the whole project is an integration of space as well as an activity of nature reconstruction. Engraved fish on the continuously outspreading white stone walls brings people feeling of freedom and lightness; white flowers like gentle breeze brings waves layer upon layer on the water surface and shows merrily and livelily digitized ink painting lotus on the surface of full body glass at the both sides of passageways. Plainly white door sheet stir up ripples ring after ring, while cobbles of various shapes adorn the space via different materials. These various natural images interpenetrate and act in concert with each other at different places of the restaurant and are integrated ingeniously by reconstruction of natural space.

All of them are selfless presents that the nature brings to us and some sprit consolation that the designer brings to the urbanites that yearn for the nature.

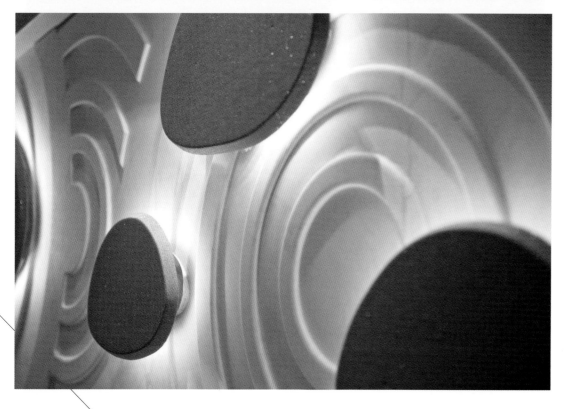

小川训央（OGAWA NORIO）

1963年 出生于日本大阪
1986年 毕业于大阪艺术大学
1989年 加入日本infix
2003年 就任上海infix总经理
2005年 就任上海infix董事长,总经理
2007年 成立北京事务所
2008年 成立香港事务所
2011年 成立上海泷屋装饰设计有限公司（SHANG-HAI RID Co., Ltd.）

Beijing Longtan Lake Paramount Chamber

龙潭湖九五书院

设计单位：Shanghai RID Co., Ltd.＼上海泷屋装饰设计有限公司
设 计 师：小川训央
参与设计：佐佐木力
项目地点：北京市崇文区
建筑面积：2650 m²
摄 影 师：贾方

本次设计，业主十分强调"古典中国的印象"。首先，就中国的历史性建筑物来说，主要的设计有独特的石阶，天花和墙面装饰。采用"在某个部分配合某种装饰"作为设想的基础，然后以此为范本，通过变换展现方式和素材以及其使用方法等进行设计。

2650 m² 宽阔的会馆内，几乎由包厢组成，风格多样的设计，能够在不同的包厢享受到不一样的氛围。比如说，天花上吊有艺术品的房间，强调突出天花的高度，通过改变很久以前就存在的中国传统艺术品的活用方式，来塑造室内氛围的变化。除此之外，通过在天花上贴木材等手法，避免"纯粹古典中式"取向的同时，还要特别注意不能设计成现代中式风格。就我们的设计初衷而言，是在"古典"这一设计基调中，同时感受到"当代"特有的时代感。会馆内，也设置了可以摆放古董的展示空间，可以享受到的不仅是店铺的设计，同时还有中国的古典艺术。

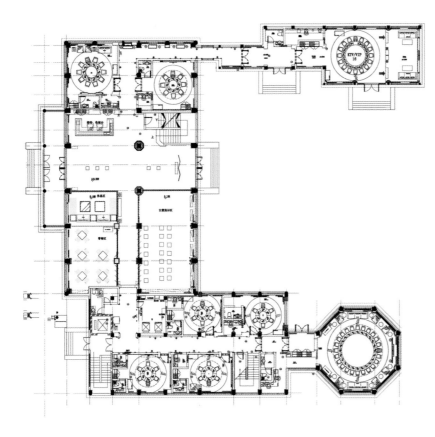

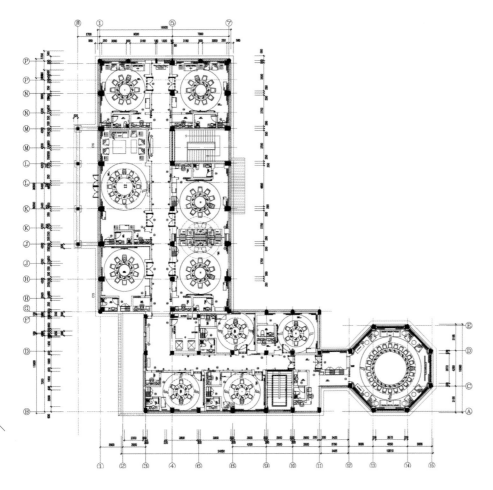

Proprietor of the project emphasizes greatly on "Impression of Classical China", which is a great challenge to us foreigners. However, I thought that it is just because that we are foreigners, we are able to display new and different beauty. Firstly, for Chinese historical constructions, unique stone steps, ceilings and wall decorations are their main design. Therefore, we base on the design concept of "Decorate Certain Parts with Certain Ornaments", regard it as the example and design by conversional displaying and through materials and their application methods…

    Almost the whole wide club house of 2650 m² is composed by compartments and design of various styles enables guests experience differently in different compartments. For example, compartments whose ceiling hangs works of art emphasizes on and highlights height of the ceiling. The change of interior atmosphere is achieved by flexible use of traditional Chinese works of art which has been existing for quite a long time. Besides, the action of pasting timber on the ceiling avoids the orientation of "Pure Classical Chinese Style"; however, special attention should be paid to avoiding designing into modern Chinese style. Our original designing intention is to enable guests feel the specific sense of "the present age" in "Classical" designing mood. There is space for displaying antiques in the club house, also. What customers of the club house can enjoy is not only the design of the club house, but also Chinese classical art.

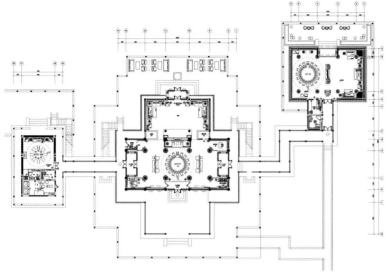

02 一层家具平面布置图@龙吟阁(比例1:100)

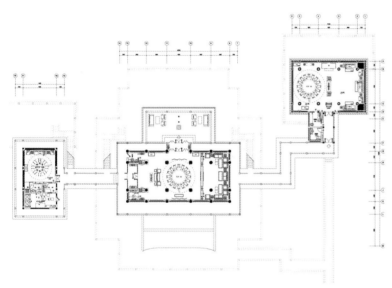

02 二层家具平面布置图@龙吟阁(比例1:100)

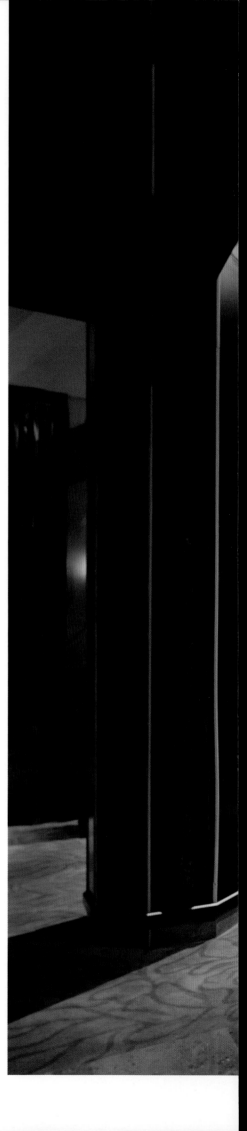

215

利旭恒

出生于中国台湾
英国伦敦艺术大学荣誉学士
古鲁奇公司设计总监

Spice Spirit Restaurant
麻辣诱惑
上海虹口龙之梦店

设计单位：古鲁奇建筑咨询(北京)有限公司
设 计 师：利旭恒
参与设计：赵爽、郑雅楠、季雯
项目地点：上海虹口龙之梦
建筑面积：850 m²
竣工时间：2012
摄 影 师：孙翔宇

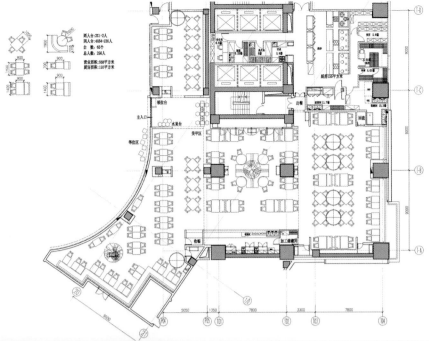

麻辣诱惑位于上海虹口龙之梦，品牌期望将女性体态曲线柔美的移植到空间概念，设计师利旭恒考虑移植原有品牌语汇的同时加入了中国太极的概念，即是在原有阴柔的基础上融入阳刚的多角砖堆砌，利用太极阴阳虚实的关系，寻找一种堆砌与互补的秩序及空间填充的概念。

餐厅用餐空间分割回三个区域，两个元素，一是白色曲线板，板子之间的镜子为了强调太极"阴"与"虚"的女性概念，曲线的构成来自麻辣诱惑品牌LOGO。设计师运用现代的手法演绎品牌形象女性曲线的基本结构，墙与顶面大量的曲线强调了人体美学，横面的曲线来自纵向曲线的迭层，借此强调女性躯体曲线之美，结构上不可置信的简单和自然。

另一元素是紫色多角砖的堆砌，密实的堆砌与大体量表现的是太极"阳"与"实"的男性概念。密实的多角体墙面纵向围合成为中心用餐区域，同时将餐厅分为三个用餐区，这一做法使得扮演男女的主题元素彼此交融，密不可分。

设计团队希望以中国传统太极概念作为基础，使用当代时尚元素具象的表现方式，让宾客在优雅的空间里享受"麻辣"的同时感受到餐饮品牌思想的"诱惑"精髓。

Spice Spirit Restaurant is situated in Dragon Dream Building, Hongkou District, Shanghai. The brand hopes to copy the softness and curved figure of woman's body into the restaurant. While keeping this in mind, Lee Hsuheng also adds Chinese Tai Chi concept into the design, i.e. while reserving the feminine and soft perception, using stacked polygonal-brick wall to present the masculine and firm sense. In this way, he takes the relations of combining masculine and feminine strength in Tai Chi to find the order of piling up and the concept of complementation and spatial filling.

The Restaurant has three dining areas which are decorated in two elements. One of the elements is the white French curves. The mirrors between boards stress the idea of feminine and softness in Tai Chi, and the shape of curves comes from the LOGO of Spice Spirit Restaurant. Lee Hsuheng interprets the brand image of curved beauty of woman's figure with modern techniques: lots of curves used on walls and ceilings highlight the human aesthetics; the laminations created by the vertical curves al-

so produce the curves at the horizontal surface, which makes the unbelievable simplicity and naturalness of woman-figure oriented structure.

The other element is the purple stacked polygonal-brick wall, the dense connection of bricks and the large volume of wall embody the concept of masculine and firmness of Tai Chi. Vertically the wall encloses a central dining volume, and at the same time, divides the Restaurant into three dining areas. This practice finishes the combination of feminine and masculine elements.

The design team wishes to take Chinese Tai Chi as the basic concept and presents it with contemporary design elements and methods, making the customers feel the true spirit of "temptation" while enjoying the delicious "spicy food"

李向宁(Li Xiangning)
意大利米兰理工大学国际室内设计硕士
经典国际设计机构(亚洲)有限公司艺术指导
北京至尚经典装饰设计有限公司艺术指导
中国建筑学会室内设计分会会员

王砚晨(Wang Yanchen)
毕业于中国西安美术学院
意大利米兰理工大学国际室内设计硕士
经典国际设计机构(亚洲)有限公司首席设计总监
北京至尚经典装饰设计有限公司首席设计总监
中国建筑学会室内设计分会会员

Meizhou Dongpo Restaurant, Yizhuang Store

眉州东坡酒楼亦庄店

设计单位：经典国际设计机构(亚洲)有限公司
设 计 师：李向宁、王砚晨
项目地点：北京亦庄
建筑面积：1700 m²
主要材料：棕云石、银白龙、真丝壁布、印刷丝绸玻璃、激光切割铁板屏风
竣工日期：2012.07

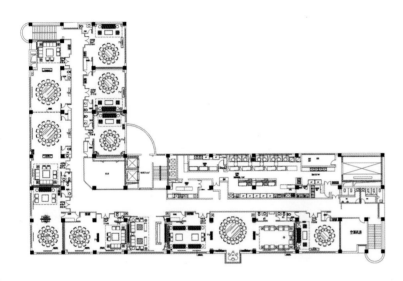

四层平面布置图

眉州东坡亦庄贵宾区项目位于北京亦庄眉州东坡的四层,是基于现已极度饱和的就餐空间新增加的楼层。1700 m²的整层面积只规划了9个包间,最大的包间面积达180 m²,最小的也将近60 m²。品牌创立人期望新的楼层能将博大的东坡文化融入奢侈的空间环境,让宾客在舒适优雅的空间里享用美食同时感受到东坡文化的浓厚氛围。

设计师尝试运用现代的手法,演绎中国传统文化的内在精神本质,展示新材料和传统材料的无限表现才能。利用水墨方式呈现中国传统哲学的处事之道,空间中充满灵动的自然之美。空间中运用大量的传统丝绸面料,与新的玻璃材质相结合,创造出充满自然美又兼顾时代感的材料,优雅地表现传统艺术的温婉雅致。

在四层电梯的入口处,一面仿古铜镜映衬出的画面着实令人叹为观止,出口对面的铁艺雕刻屏风巧妙隔开等候休息区和收银区,传统的中式屏风,由最现代的钢铁材料和激光技术重新演绎,呈现出迥然不同的审美情趣,使空间像充满禅意的中式园林。

空间中大幅水墨山水壁画的载体不再是传统的宣纸和绢,而是丝绸玻璃,东坡泛舟赤壁的经典画面也被重新解构,山水和人物分为两层,在玻璃和丝绸的映衬下,随着视线的移动,山水和人物在空间中形成新的视觉印象。这些层迭变化的界面将内部空间进行了重新定义,顺应了空间跨度的美学需要,并传递了某种微妙的动感。

Located at the forth floor of Beijing Yizhuang Meizou Dongpo restaurant, the project of Meizou Dongpo Yizhuang distinguished guest area is a new added floor based on the extreme saturated repast space now. Only 9 rooms of planning on 1700m2 area of the whole floor ,the area of the maximal room is 180m2 and almost 60m2 of the minimal. The founder of the brand expected that the new restaurant could integrate the vast Dongpo culture into the luxurious spatial environment, offering strong atmosphere of dongpo culture to customers while enjoying the delicacy in the comfortable and elegant room.

Designers had tried to apply the modern skill to perform the inherent spiritual essence in Chinese traditional culture, showing the unlimited performance ability to new and traditional material. Utilization of ink painting to present the way of the philosophy of Chinese traditional dealing with affairs, fulfilling the flexible natural beauty in the room. Integrated the utilization of much traditional silk plus material in the room into new glass material, which will create the period feel material fulfilling natural beauty and elegantly express the gentle and tasteful traditional art.

A mirror in the style of the ancient silhouette against a screen which acclaim as the peak of perfection on the entrance to the elevator of fourth floor. Located at the opposite to the exit, the screen graved in steel art delicately separate the await rest area from cash area.

The traditional Chinese screen is performed again by both the most modern steel material and laser technique, presentation on the widely different aesthetics temperament and interest, which make the room being like the Chinese style gardens fulfilling zen's real meaning.

The carrier of the substantially ink landscape wall painting in the room is no longer traditional rice paper and tough silk but silk and glass. The classic picture of Dongpo yachting around Chibi had been destructed again, dividing landscape and characters into two layers. Against the background both in the silk and glass, landscape and characters had shaped new vision image with the moving of sight line. These interfaces varied in stacks had redefined interior room. Complying with the aesthetics requirements to the spacing of room and delivering some delicate sense of movement, which place yourself not in static room any longer.

朱晓鸣

毕业于浙江树人大学
杭州意内雅建筑装饰设计有限公司 创意总监/执行董事

Mango Thai in Ningbo
宁波美泰泰国餐厅

设计单位：杭州意内雅建筑装饰设计有限公司
设 计 师：朱晓鸣
参与设计：张天明
项目地点：浙江省宁波市
建筑面积：300 m²
主要材料：定制仿古花片砖、石灰稻草泥、老木板、橡木染色、密度板雕刻、陶土瓦、户外地板、编织木
摄 影 师：林峰

此案位于商场三层,面积并不大,外优势并不明显,注重打造空间的内优势,借以独特的自我特征识别,达到良好的传播是我们设计中考虑的重点。本案在区域划分中反常规地将餐厅入口移置到人流交通的最远端,使来访者在迂回的踱步中对整个用餐氛围有所感知,刻意使客户在餐厅门口廊道中有所积流而不是快速分流,进入等待区后,再通过双通道分流。用餐区根据人群结构不同,割划了对坐区、卡区、散座区、包厢区等。条卡区可酌用餐人数快速拼接来改变其接纳量。而泰式礼品区、水吧区、收银区等整合式"中岛"设计,既减低工作人员的数量,又在视觉交叉点上有极佳全景视线,增加了服务的快速便捷性。包厢外走廊的过渡空间划分,使其区域有着远离大厅用餐区的视觉错觉感,更加静谧、独立。在设计风格的导入中,考虑其建筑的层高,及结合来访者的年龄层特质,并未一味地将泰式暹罗建筑特质的灿烂辉煌、塔尖翘角等较为异域华丽的元素强加运用,而是在整体较为质朴、平和、随性的"底色"中略施粉黛,恰当地加入了有泰式民族特色的鲜活图案及色块,进行"矛盾"的破坏,如地毯式定制花砖铺设,阵列的多彩门框,绮丽的家具面料运用增加空间的视觉张力;大厅区、包厢区叠水瓦墙的延续运用加以热带植物的烘托,增加其热带南国的风情。借此打破传统泰式风格的贵族仪式感与单调沉闷。空间既有泰国风情又有再创的现代时尚感,创造了充满热带气息的轻松用餐环境。

The restaurant is located on the third floor of the shopping mall and with not large area and comparatively ordinary external advantages. For these reasons, creating the restaurant's internal space advantages and achieving good promotion effect by the restaurant's unique features becomes the important points that should be considered. This case does not follow the conventional regional division, but displaces the entrance of the restaurant to the most remote terminal of stream of people so as to enable customers to have a concrete feeling of the whole dining atmosphere in roundabout entering and intentionally creates the situation of customer gathering at the gate corridor of the restaurant instead of splitting flow quickly. After customers' stepping into the waiting areas, the stream of customers will be split by twin channels. According to the difference of the crowd, the dining area is divided into area for two people, area for group, area for odd customers, compartment area and the like. Area for group may be modified quickly to provide suitable accommodation capacity by jointing. And the integrated design "Middle Island" which integrates Thai-style gift area, restaurant and diner area, cashier area and the like not only help reduces staff amount, but also form excellent full view at the intersection of vision and improves the speed and convenience of the service. The division of transitional space of corridors outside the compartments creates a visual illusion that this area is far away from the dining area and leads to a feeling of serenity and independence. While in the aspect of design style, the designer does not blindly strengthen the application of the comparatively luxuriant features of Thai-style Siam constructions, such as its magnificence, spire & rake angles and the like, but take floor height of the construction and age situation of customers into consideration. The designer gently decorates the construction in its comparatively plain, peaceful and casual "grounding" and suitably adds fresh and alive patterns and color lumps with Thai national features to carry out "contradiction" violation, for example, use customized carpet-style tiling, colorful array doorframe and gorgeous furniture fabric to improve the visual tension of the space. The continuous use of small cascade and tile walls at the hall area and compartments area are decorated with tropical plants to increase its tropical austral amorous feelings and break Thai-style's sense of noble ceremony and monotonous depression. The space not only has Thai amorous feelings, but also recreated sense of modern fashion and light dining environment with tropical atmosphere.

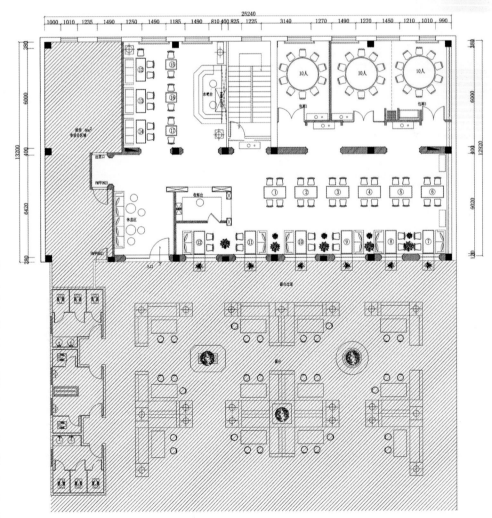

一层平面图
Scale 1:80

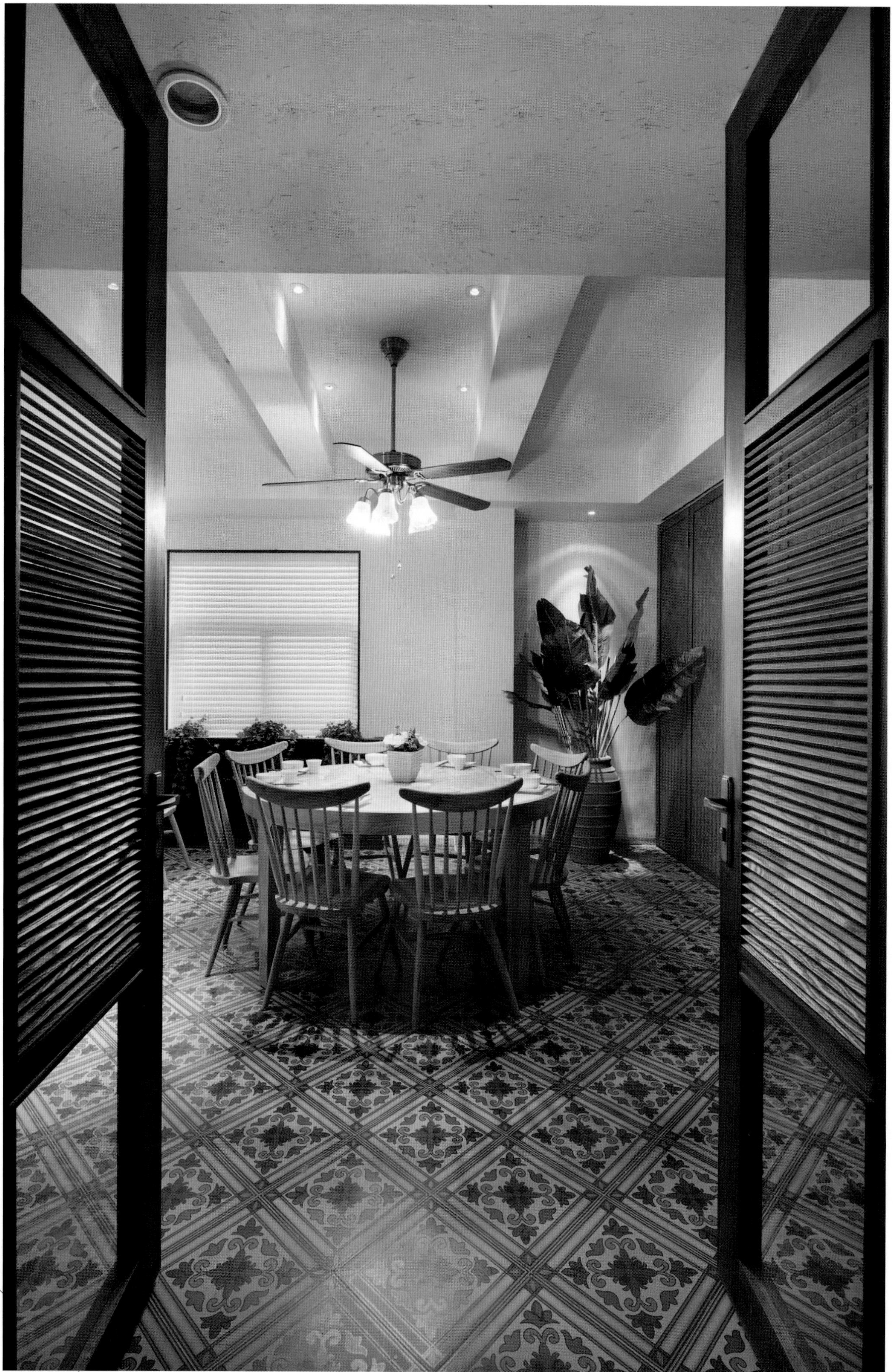

熊华阳

深圳华空间机构总经理\设计总监
中国建筑学会室内设计学会会员
高级室内建筑师

Club House of Honor, Tongren, Guizhou

贵州铜仁上座会馆

设计单位：深圳华空间机构
设 计 师：熊华阳
项目地点：贵州
建筑面积：7000 m²
主要材料：大理石，绿可木，布艺沙发，装饰画

坐落于贵州省铜仁市的上座会馆,是接待当地高端人群及高级客商的重要场所。在这样一座有山有水的自然静寂之城,最适合于它的建筑莫过于传承我们几千年历史文化的中式风格建筑了。上座会馆由两座三层高的中式建筑砌合而成,由建筑外观到室内设计,及软装配饰等,均由华空间一体化完成。会所内含有健身会所、娱乐酒吧、中式餐饮等项目。

  设计师在设计项目时需扬长避短,充分利用项目的优势,上座会馆依托于当地的山水之景及少数民族特色,所以从外观设计到室内设计,使用中式的设计框架结合现代风格的家具、饰品,院子中央的小池塘、少数民族特色的壁画、现代风格的沙发、中式古典的木椅、竹叶图案的地毯……使会所由内及外散发出传统的高雅的新中式设计风格。

  空间的设计不在于将墙面及吊顶做复杂的处理,而在于给它恰到好处的点缀;古典中式的沙发是否会有些呆板?我们结合现代简约的沙发一同陈设,流畅的线条即显现出来;普通的过道如何才能拥有设计感?我们使用镜面并带有图案的玻璃做墙面;正统的中式包房拥有古典的中式家具就足够了吗?我们设计了与众不同的玄关增加了包房的设计感;您是否忽视了楼梯的设计?但我们注重项目的每一处细节,选用几何形状的时尚楼梯扶手,并在楼道陈列着艺术品展示……

  无论是室内设计,还是产品设计,成功的设计在于相辅相成,由点及面的互相呼应。如此案中多处应用的方圆结合,会馆外观,接待大厅,池塘边上,包房内的玄关,都是方圆之间的艺术组合。

Club House of Honor which is located in Tongren City, Guizhou Province is an important place for receiving local high-end groups and high class guests and businessmen. To a natural and quiet city with both mountains and water, the most suitable construction for it is Chinese style construction that inherits our historical cultures of thousands of years. The Club House of Honor consists of two three-floor Chinese style constructions. All things of the club house, from the fa?ade of the constructions to their interior design, soft decoration & ornaments, etc., are completed comprehensively by Hwayon. The club house contains items such as body building club house, entertainment bar, Chinese style repast…

While designing for projects, designers should adopt the project's good points and avoid their shortcomings and take full advantage of the projects. Relying on local landscape and minorities' national features, the Club House of Honor integrates Chinese style design framework with modern style furniture, decorations, small pond in center yard, murals with national features of minorities, modern style sofa, wooden classical Chinese style chairs, carpets with bamboo leaf patterns from the design of fa?ade to interior design… Therefore, the club house is able to express fully traditional and elegant new Chinese style design style from the interior to the exterior.

Space design does not lie on complex treatment to wall surface or suspended ceiling, but lie on ornament to the point. Will classical Chinese style sofa seem a bit rigid? Display it along with modern simple sofa will highlight flowing lines. How to make ordinary passages with graphic sense? Let us make mirror planes and glass with pictures the wall surfaces to settle the problem. Will it be enough to equip orthodox Chinese style compartments with classical Chinese style furniture only? The creating of extraordinary hallway will enhance the graphic sense of the compartments. And have you ignored the design of staircase? We pay attention to every detail of the project and for staircase we choose fashionable geometrical shape stair railing and display works of art at the passageway…

No matter it is for interior design or product design, the successful ones should be those whose parts can well supplement each other and whose points and surfaces take concerted actions, such as this case which integrates square and round in many places: façade of the club house, the reception hall, pond side, hall ways in compartments… All of them are art combination of square and round.

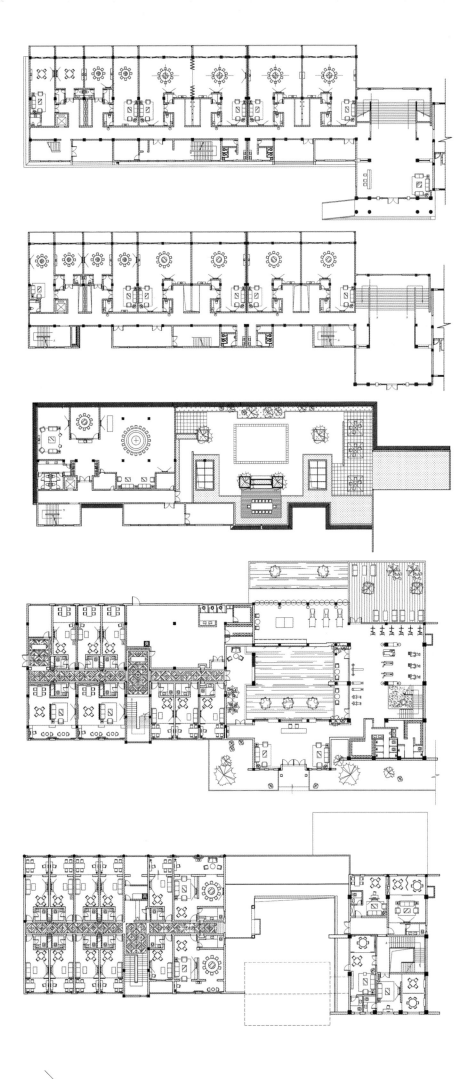

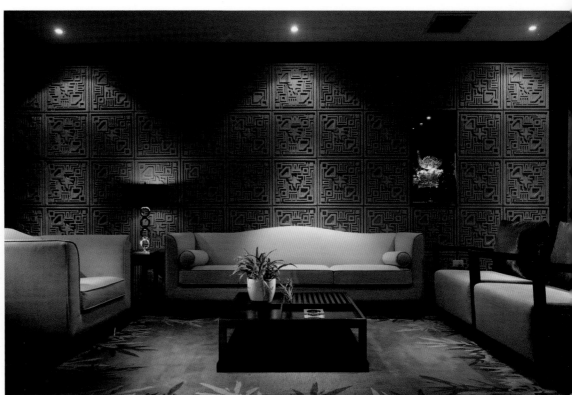

陈彬

武汉理工大学艺术与设计学院副教授\硕士生导师
CAAN 中国美术家协会会员
CBDA 中国建筑装饰协会设计委员会委员
CIID 中国建筑学会室内设计分会会员
IAI 亚太建筑师与室内设计师联盟理事
ICIAD 国际室内建筑师与设计师理事会会员
大木设计中国理事会常务理事
后象设计师事务所创始人\设计主持

Design Description of the Favorite Pavilion

所好轩（沌口店）

设计单位：后象设计师事务所
设 计 师：陈彬
项目地点：武汉
建筑面积：360 m²
竣工时间：2011.12
摄 影 师：吴辉

所好轩（沌口店）是主题店中店的再次演义，设计师依旧将色彩选定为空间主角，继续延续红与蓝的主旋律，在灯光的营造下，上演了一出空间梦幻剧。

特别定制的漆面仿纸折皱钢板天花，从网速缓慢而被延伸的图片中获得灵感的红色透隔，点缀着传统道家星宿图的丝绒皮制墙板，模仿自然晨雾和雨丝的特效隐光玻璃被巧妙地组合在一起，使空间充满戏剧性，并传达着设计师以这个时代特定的方式向传统文化的敬意。

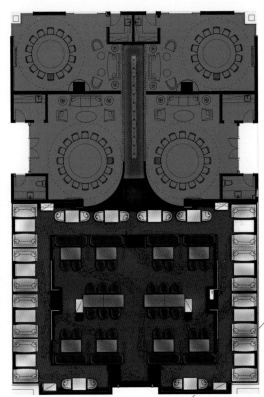

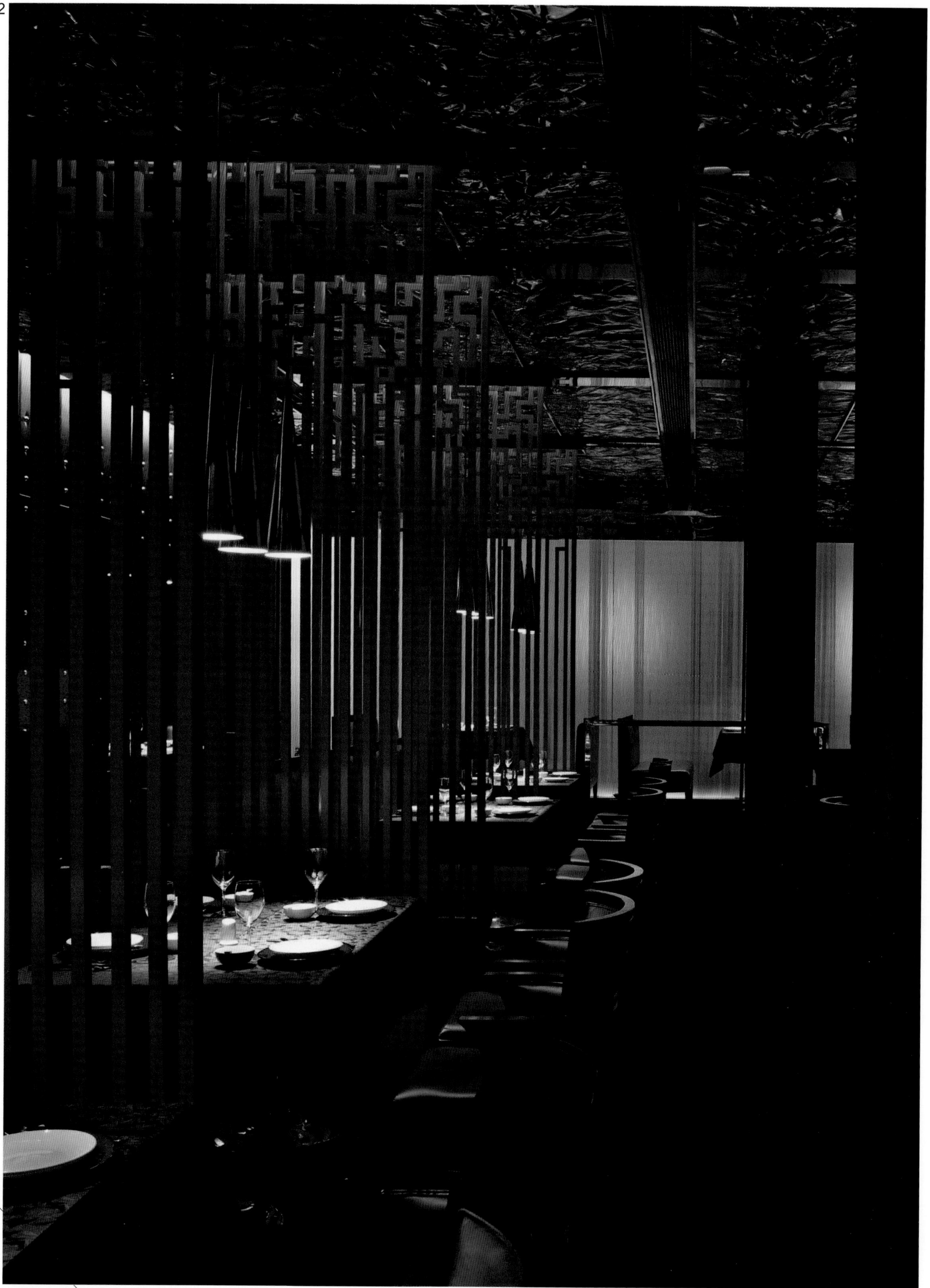

Favorite Pavilion (Zhuankou) is another demonstration of the themed restaurant in a main restaurant. The designer still chooses colors as the major role of the space with an extension of the theme of red and blue color. The lighting effect creates an atmosphere of a spatial dream play.

The artistic combination of customized ceiling of synthetic lacquered paper and wrinkled steel plate, the red transparent heat insulation inspired by the extended pictures due to slow network speed, the velvet leather wall board dotted with the constellation drawings of the traditional Taoism and the frosted glass imitating the effect of natural morning mist and drizzle creates the dramatic feature of the space and expresses the designer's respect to the traditional culture in a specific way of the contemporary era.

王奕文
和合堂设计品牌创始人\设计总监
蓝犀装饰公司设计总监

Bianyi workshop
便宜坊

设计单位：和合堂设计咨询
设 计 师：王奕文
参与设计：赵光宇、胡岩峰、吴林国
项目地点：北京华润五彩城
建筑面积：1400 m²
主要材料：建筑灰砖、木雕、壁纸、金漆、镜框、绘画作品
摄 影 师：毛立广

　　灵感来源于雍容华贵、富丽端庄的牡丹，本案是和合堂设计在"便宜坊"连锁店中的又一创新，寄予其牡丹花卉图案雕刻、牡丹颜色寓意等手法，运用布局、色调、灯光、材质等来演绎富贵吉祥、繁荣兴旺的氛围。

　　入口8 m挑空位置为富丽大气的金色门面提供了不可多得的空间条件，弧形收银台的二龙戏珠实木雕刻，背景8 m的木隔断，红色写意绘画作品，无处不渗透着高贵典雅端庄秀丽的视觉冲击力。

　　与入口相邻的区域为半开敞包间区，设计师赋予其牡丹花瓣般的平面造型演绎，完成空间的半封闭状态，灰色轻纱飘渺间，红色灯笼造型吊灯若隐若现，仿佛亭台楼阁间的婆娑倩影，令用餐空间变得更为趣味性和戏剧性。

　　顺着动线的指引豁然开朗，大的散座区映入眼帘，天花延续了花瓣圆形造型，此区域可容纳200人同时就餐，为婚宴、大型聚会提供场所。

　　二层是浓墨重彩的包间区，狭长的弧形走廊，一侧为歌颂便宜坊历史的金色词句雕刻，一侧为以牡丹的色彩为线索而设计的五个包间，黄色、红色、绿色、白色、墨色的半开敞空间将略显深邃的通道变得情趣盎然。

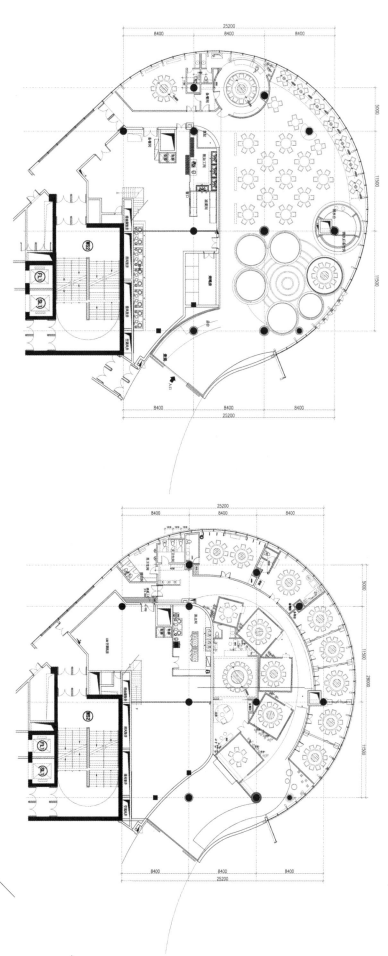

With the inspiration from peonies both of amazing grace and wealthy as well as elegant, this case is another innovation from the design of chain store "Pianyi workshop" by Hehe hall, which places on techniques such as sculptures of peony floral prints and implied meaning of peony color, and applies layout, tone, lighting, material qualities etc to present the atmosphere of wealth and rank as well as prosperous.

With golden appearance both of wealthy and meteoric in the space which is 8 meters from the entrance, which had provided rare spatial terms that there permeates visual impact of noble and elegant as well as dignified and beautiful everywhere such as solid wood being carved with 2 dragons playing with a pearl on arc cashier desk, a background of 8 meters' wood obstruction, red enjoyable paints.

With semi-open type private rooms neighboring the entrance, designers had given plane shaping presentation like peony leaves to it to complete semi-enclosed status. Red lanterns shaping droplights are partly hidden and partly visible among dimly discernible gray fine gauze as if dancing pretty images among airy pavilions and pagodas, having made dining space being more interesting and dramatic.

Large seating area for odd will come into view at the moment of being suddenly enlightened along with the guidance of movements, having a capacity of 200 people to have meals at the same time and provided a scarce place for wedding party and large gathering.

There are private rooms with thick and heavy in colors on the second floor, on both sides of the long and narrow arc corridor, one is golden expressions which eulogize the history of Pianyi workshop, and the other is five private rooms which are designed as clues to peony color, the semi-open space of yellow, red, green, white, ink color, which will make the slightly deep corridor become temperament and interest.

The case perfectly integrates both qualities connotation restraining & casual nature by simply Chinese method. Chinese brief style is applied interspersed in the design and demonstrates beauty that incorporates things of diverse natures instead of showing improperness. Convert the so-called classical words by geometrization, visualizing, comparison or rhythmization. Change "complex" ones into "simple" ones in the aspect of lines, emphasize on simple color that is dignified and worshipful in the aspect of tone and propose decorousness, texture, luxuriousness or even "coming straight to the point" in the aspect of space language communication.

The Chinese garden style technique used on facades refreshes peoples at their first glances. The dealing result is imposing on the whole. And the entrance applies screen wall facing the entrance and the screen expresses fully the artistic conceptions of winding path's leading to a secluded quite place and landscape changing by step moving. The interior design is under good control also. The whole rhyme, material and color are under full control and elements of Twelve Girls Band are used vividly.

Based on Chinese native culture, the designer perfectly integrates Hui-culture and Chinese culture by using a variety of patterns of manifestation. Some designed parts are edited and arranged Hui-culture. The designer uses Orchid Pavilion to show the design conception of the case. The designer put forth effort to break the design status in this case and find the innovative design style that is suitable for Chinese and enable people understand truly the essences, such as conciseness, briefness and refinedness, within design of Chinese and constructs space that is with poet's feelings. Within the design of some compartments, dining area and relaxing area are reasonably separated and really high courtyard is equipped in the relaxing area.

张健

毕业于杭州某美术类学校
1999年起从事室内设计工作
2007年底创立杭州观堂设计

Little Cook Seafood Restaurant
小厨师海鲜餐厅

设计单位：杭州观堂设计
设 计 师：张健
参与设计：陈蕾
项目地点：杭州保俶路
建筑面积：160 m²
主要材料：旧木板、旧门板、瓦片、石材
摄 影 师：王飞

餐厅以新鲜、美味的特色海鲜为主,在设计中着重运用了海边元素,门头采用大块的石材,营造海边礁石的氛围;一层左侧外墙采用各种瓦片、砖块排列成不同的图案组合,一来外观吸引眼球,二来形成半通透的墙面隔断,路人可以看到室内大厨们忙活的身影,也为厨房的通风提供更好的条件。

餐厅室内设计中,采用回收的砖块、门板、窗花、木箱铺就墙面,营造一种随意、放松、粗犷又不失细节的就餐氛围,仿佛海边小镇上的一家小餐厅。餐桌、餐椅、木梁、窗框、隔断、楼梯、扶手也都采用回收的老木板,环保又温馨。

由于餐厅本身空间有限,因此在隔断、窗户上面尽量做得通透,临街窗户全部采用透明玻璃,并且尽最大可能的打开,让客人在就餐的同时,能随时与外界新鲜的空气、郁郁葱葱的绿植互动。

卫生间与洗手盆的墙面地面则采用破碎的陶瓷片装饰,如咖啡杯、马克杯、陶碗等的碎片,五彩斑斓,形状各异,让客人在如厕同时欣赏不一样的风景。

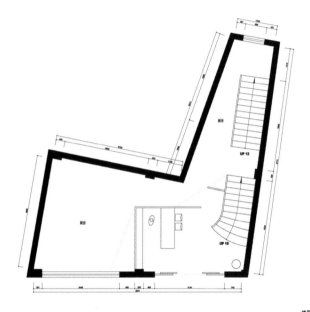

一楼平面布置图 1:50

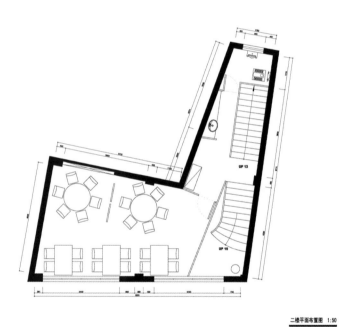

二楼平面布置图 1:50

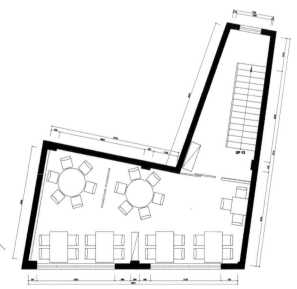

三楼平面布置图 1:50

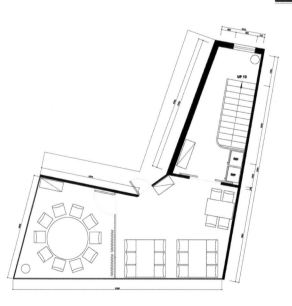

四楼平面布置图 1:50

The restaurant gives priority to fresh and delicious characteristic seafood, applies seaside elements during design, uses massive stone at the top of doors to build the atmosphere of seaside rocks; the left to a bed uses all kinds of tiles and bricks to arrange into different pattern combinations to attract eyeballs by its appearance on one hand, on the other hand, to form semi-permeable wall space obstruction so that not only passerby could see busy figures of chefs indoors but also provide better conditions for airy kitchen.

During the indoor design to the restaurant, uses recyclable bricks, door sheets, paper-cut for window decoration, wooden box to pave wall space to build repast atmosphere of being at will, relaxed, rough without losing details as if a little restaurant on seaside of a town. Dining-table, dining chairs, beams, window frame, obstructions, stairs, handrails are all adopted old board, which are both environmental protective and sweet.

Due to the limited space of the restaurant itself that make both obstructions and windows being permeable to the greatest extent, whole use transparent glasses in the frontage windows and could open at the utmost to enable guests to interact both with fresh air outside and luxuriantly green plants at all times at the same time of repast.

Use crush ceramic chips which are both colorful and different shaped such as coffee cups, mugs and clay bowls to decorate both wall space and ground both of rest rooms and basins, and enable guests to appreciate diverse scenery at rest room.

潘开富

毕业于南京林业大学

南京智点室内设计顾问有限公司设计总监

Tongqing Building Luzhou Fu
同庆楼庐州府

设计单位：南京智点室内设计顾问有限公司

设 计 师：潘开富

项目地点：合肥

建筑面积：20000 m²

摄 影 师：吴辉

同庆楼庐州府坐落在合肥市马鞍山路与东流路交口,大厅入口处景观,高梁直柱,徽派之风彰显大气。酒店以徽派园林设计风格为主,或时尚典雅,或恢宏华丽。宴会大厅气派非凡,可承接近200桌各类宴席,配备顶级音响设备、LED显示屏,甚至可以根据顾客需要任意分割空间。在硬件装修选材上用的是进口名贵石材和名师设计的灯具,随便一件便数万甚至百万,为顾客打造最舒适最有质感的就餐空间。

Tongqing Building Luchou Fu is located in the cross of Maanshan Road and Dongliu Road of Hefei City. At the entrance of hall, you can see high beam and right cylinder, style of Huipai and elegance. The hotel is characterized by garden design style of Huipai, fashionable elegance or magnificent and gorgeous style. The banquet hall is extraordinary and can accommodate about 200 tables of all kinds feast, equip with top-level audio equipment, LED display, and can even divide space arbitrarily according to customer needs. As for hardware decoration materials, it selects imported precious stone and lamps and lanterns designed by famous designer. A simple decoration article is with ten thousands or even several millions in order to create the most comfortable and most textured dining space for customers.

宋国梁

中国最具影响力室内设计师
中国百名优秀室内建筑师
ICIAD 国际室内建筑师与设计师理事会宁波区理事
IFI 国际室内建筑师设计师联盟会员
CIID 中国建筑学会室内设计分会会员
新加坡 V.特锐建设集团/总设计师

Yongxiang Fashionable Restaurant
涌香格调时尚餐厅

设计单位：新加坡 V.特锐建设集团
设 计 师：David/宋国梁
参与设计：M.G 设计组、N.O 设计组、纳海川配饰组
项目地点：宁波市鄞州区樟溪北路88号
建筑面积：2300 m²
主要材料：木皮、墙纸、镜面不锈钢、大理石
摄 影 师：刘鹰
撰　　文：野风

涌香格调时尚餐厅地处中国宁波美丽的奉化江畔,位于商业热土舟宿夜江中心地块,环境幽静恬淡。 精彩纷呈的时装发布会成为设计师创作的灵感,流动的色彩、混搭的文化元素与本案有着异曲同工之妙,品味、气度、涵养如描绘年轻绅士般写满了从容。 餐厅融合现代、西式复古、时尚元素,无论北欧风情、田园诗意、现代情愫,都显得巧妙流畅。整个餐厅洋溢着浓浓的文化气息,书卷满廊、原创前卫艺术装置、手工陶盘、名师画作,从内到外透着优雅,造就了餐厅的特有气质。 从一到三楼贯穿的共享空间,为原有楼梯间改造,斜洒阳光、清流潺潺、灵动、活泼、充满生机。包厢风格迥异,各具特色。三层多功能户外餐吧,坐拥甬城绝美夜色,赏佳景、享受美味佳肴、浪漫情怀,为甬城独一无二的特色场所,也是人们沙龙、party的理想去处。 不管是家具、灯具、特色工艺装饰,每一个细节都经过设计师精雕细琢,设计师本人创作的装置艺术给空间增添了独特的魅力。

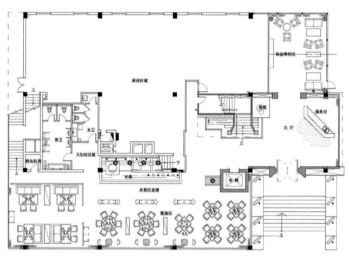

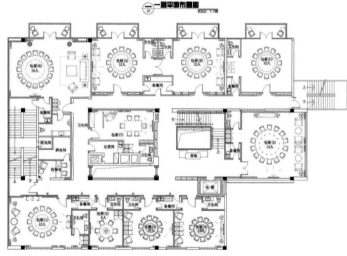
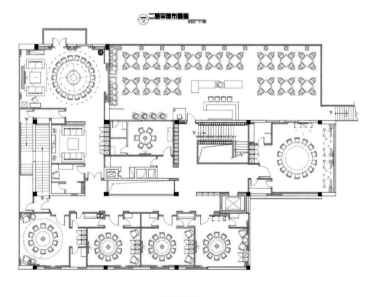

By the side of beautiful Fenghua River, Ningbo, China and being located in the central commercial place, Zhousu Yejiang, Yongxiang Fashionable Restaurant is with quiet and tranquil environment. Colorful fashion show, whose flow of color and mix of cultural elements are different from this design in approaches but equally satisfactory in results, became designers' creative inspiration. The restaurant's demonstration of good taste, bearing and cultivation is shown as those of young unhurried gentlemen. The restaurant integrates modern style, Western vintage style and fashionable elements. No matter it is Nordic style, pastoral poetry style or modern sincerity, the atmosphere is created elaborately and smoothly. The whole restaurant runs over with strong cultural atmosphere; a gallery will full books, original avant-garde art devices, handmade pottery plates and masters' paintings create the unique features of the restaurant and demonstrate a sense of elegance from the interior to the exterior. The shared space connecting the first, the second and the third floors is transformed from original staircase. Oblique sunshine shines, clean water murmurs and creates active and vital atmosphere in the space. And compartments of the restaurant are in different styles and with various features. Three floors of multifunctional outdoor bars provide excellent places for appreciating beautiful night views of Ningbo. That good views, delicious food and romantic themes may be available here makes Yongxiang Restaurant an unmatched characteristic place in Ningbo and also an ideal place for salon and party. Every detail, no matter it is the detail on furniture, on lamp, or on specific artistic decoration, it has been elaborately refined by designers; the art installation created by the designer in person adds unique charm to the space.

周易

- 1959年 出生于台中市
- 1979年 以自学方式学习建筑与室内设计
- 1989年 创设JOY室内设计工作室
- 1995年 创设JOY概念建筑工作室

Co-exist & Harmonious

客家本色大里店

设计单位：周易设计工作室
设 计 师：周易
项目地点：台湾台中市
建筑面积：895 m²
主要材料：铁件、清水模、枕木、玻璃、大理石、抛光石英砖、竹子
竣工时间：2011.09
摄 影 师：何风摄影

这栋上、下仅两层楼高的独立建筑,清水模的质朴、陶缸水景的写意、加上玻璃帷幕的通透感,在鲜明的建筑语言里,同时注入了业主期许的自然、休息概念,让人第一眼就留下生动印象。

走过户外粗犷的枕木栈道,单侧有着格栅窗花图腾的长列方形步道灯,以典雅的光影诉说着迎宾的热忱。栈道与主建物间规划水景区,以黑色石材砌成的无边界水池里,点缀着四座巨大黑色陶缸,以及状似漂浮的烛台灯与光束涌泉,水池中央一座强调极简线条美学的清水模结构,让室内外有了适度的衔接与屏障,设计者借由这些源于自然界的木、石、光、水等元素,传递一种人造工艺与自然共生的极致,提炼宁静与安逸的环境和谐之美,让所有到访者都能以最放松的心情入内用餐。

一进门所见即为气宇非凡的柜台区,善用建筑物局部挑高逾7m的优势,在柜台后靠端景处以工序繁复的白底浮雕隶书;加上分毫不差的铅字排印技巧,将李白一首潇洒、豪情万丈的《将进酒》,在精妙的聚焦灯光下,进化为气势万千的立体视觉艺术,而点缀文字群间;大小不一却同样别致的红色落款,灵感来自乾隆皇帝喜欢在钟情的书画上捺印,隐喻专属专有的独特性,整个画面的经营,除了意蕴其中的人文涵养,更是一次独到美学的具体实践。此外,文字端景墙前精选黑色石材打造柜台主结构,立面以错落的实木块传递不经修饰的自然感,上方一排利落铁架运用纤细的钢骨深入天花板内强化支撑,架上一长排烛灯是室内外共通的元素之一,综合以上极度情境的构图手法,千年的文化重量与休息的丰富内容尽在其中。

餐厅内部以时尚的黑白对比为主调,但造型语汇上却穿插了许多古老

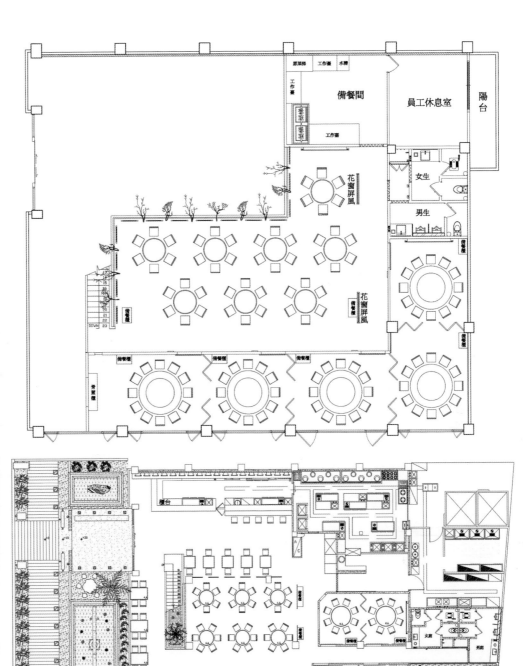

中国的片段，例如漆黑的梯间下缘，一方镜面池子仿佛是户外水景的复刻版，池中一棵全白枯枝，既有白山黑水的泼墨意境，更有日式枯山水的耐人寻味，尤其经过设计师周易一向擅长的灯光烘托，更将生活的无限美好浓缩于眼前的方寸之间。一楼主要为开放的用餐空间，其中一侧运用"有景借景，无景则避"的技巧，将落地窗外优美的竹林景致吸引入内，大大提振食欲和情绪的感染力，不容错过的还有点缀在窗畔白墙与特定包厢内的大幅书法艺术，全是当代知名书法家李峰的作品，其中一幅名为"如易"，很巧合地将设计者与业主名字中共有的"易"字带进来，营造既有象征性又意义深远的客制化艺术。另一侧包含夹层上下均使用大量中式窗花分段界定，全数喷白的线条格外立体，一字排开的气势营造"数大便是美"的震撼力，二楼天花板处还有黑色枯枝迤逦而出，串连空间处处呼应的设计主题。

　　私密包厢的设计同样饶富巧思，在隔墙上缘点缀的白色竹簧与灯光阴影，让人联想起"采菊东篱下"的悠闲，以雅致瓶门发想的入口造型，有着中式园林的书香气质，门上的厢名则以知名的客家聚落命名，这也点出了业主源出于此不忘本的初衷。

JOY Interior Design Studio makes good use of original building's congenital condition such as high interior ceiling room and floor to ceiling continuous windows and use an open field of vision and philosophical, and harmonious logical thinking to bring the Suzhou gardens like quiet indistinct of water, bamboo grove and narrow footway from external view into indoor box view, it also integrate modern classic black-white fashion contrast and Eastern culture fashion vocabulary and change and reconcile according to various proportions. Through the conflict and harmony between the two, it writes a whole new comment on the feeling of a pleasant dinner gathering.

Walk by the outdoor rough railroad tie, one side has a long line of square trail light with grid square window totem; elegant light and shadow tell the sincere welcome. Between the footway and the main building, a water view area is planned, in the borderless pool built with black stones, there are four huge black ceramic vats and candle lights and light beam springs which seem like floating on the pool are used to decorate the pol. In the center of the pool, an exposed form structure which stress minimalist aesthetic lines make appropriate connection and barrier between indoor and outdoor. Though the nature elements such as the wood, stone, light and water which were originated from the nature, the designer delivers the ultimate attainment between artificial technics and nature and refine the beauty of environmental harmony through quiet and comfortable environment.

When entering the door, one could immediately see the extraordinary counter area. The advantage of 7 meter high ceiling in some area of the building is properly utilized, in the scene behind the counter white relief official script made with complicated process and almost perfect print typography skills transform the poem made by Li Bai with chic and lofty heroic feeling under the exquisite focus light into the three-dimensional visual art with strange momentum to decorate the words; the same chic red inscription with different size, and the conduct of the whole picture, in addition to implication with human virtue, is an unique and concrete practice of aesthetics.

The interior of the restaurant is using the fashion white-black contrast as the main tone, but the styling vocabulary is interspersed with a number of fragments of ancient China, for example, the lower edge of the dark staircase, a mirror like pond just like a replica of outdoor water views. A white dead branch in the pond has not only the "splash-ink painting" artistic conception of mountain and rivers, it also has the interesting scene of Japanese mountain and water, especially after the designer-Chou Yi who is always good at utilizing lights to decorate and make the goodness of life concentrated the view in front of our

eyes. The first floor is mainly the open dining space, the beautiful scenery of bamboo forest outside the window is introduced into the interior though the French windows which greatly boost the appetite an emotional appeal.

The design of private box is also full of clever thinking, the white bamboo canopy and light shadow decorated on the edge of the wall make us associate with the laid-back feeling in a famous poem. Entrance door style is using the idea of elegant bottle; it has the sophisticated temperament of Chinese-style garden. The private boxes are named after famous Hakka community, it also points out the owner's idea of not forgetting ones origin.

高雄

道和室内设计有限公司 总经理/设计总监
资深室内设计师
中国建筑室内装饰协建筑室内设计师
中国建筑学会室内设计分会会员
建筑装饰装修工程师
IAI 国际室内建筑师与设计师理事会 华南区及福建代表处理事

Impression of Wang Jiang Nan
印象望江南

设计单位：道和设计机构
设 计 师：高雄
参与设计：郭予书
项目地点：宁德市
建筑面积：287 m²
主要材料：白色烤漆波浪板、黑镜、黑钛、枫木饰面板、白色烤漆玻璃、黑白根大理石
竣工时间：2012.08
摄 影 师：周跃东

印象望江南餐厅位于宁德万达广场内,该店的使用面积为287 m²,餐位数为122人位。

　　本案黑白色相间、简约、线条具有一种直抵人心的力量,整体设计呈现时尚、简洁的气质。

　　黑白色反差、镜面反射的场景,具有强烈的空间感。店内家具均以黑色皮革及大理石台面为主,墙面造型均以凹凸的波浪板及镜面为主,强烈的厚重感与跳跃简洁的线条给客人带来感观的冲击,也体现了餐厅的品牌定位。整体色彩对比强烈,但空间更富于流动和变化。

　　白色烤漆玻璃墙面若隐若现的山峦,给人无限遐想。仿佛置身江南水乡。

　　入口大厅的处理是本案的出彩之处,以玻璃钢雕塑倒挂或坐的各异姿态,加上中式的白色喷漆鸟笼以及栩栩如生的小鸟,时尚现代感与东方气质的碰撞,为整个餐厅增添许多各异的色彩。

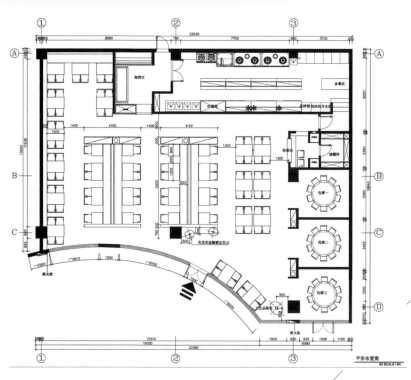

The restaurant of Impression of Wang Jiang Nan is located in Utama shopping center in Ningde, which covers area of 287 square meters and boosts dinning spaces of 122.

Black is alternating with white, being brief, and lines of this case have powers of straight through the heart, whole design is presented on temperament both of fashion and brief.

Contrasting black with white and the scene of mirror reflection have strong sense of space. The furniture in the store is mainly black leather and marble platform, modeling of wall space is mainly concavo-convex wave board and mirror surfaces, strong massive sense and lines both of jumping and brief had brought customers impacts of perception, which will also reflect brand positioning of the restaurant. Contrasting colors of the whole, but spaces are richer both in flow and change. Wall spaces of white baking finish glasses and mountains both of partly hidden and partly visible had brought infinite reveries, as if placing oneself in Yangtze River delta.

The treatment of the entrance to the hall is distinctive, different gestures of sitting or dropping away of sculpture of glasses reinforced plastic, and white spray painted birdcage of Chinese style with lively birds, sense both of fashion and modern crashes eastern temperament, which had added much different color to whole restaurant.

潘鸿彬
泛纳设计事务所创始人
香港理工大学设计学院助理教授
香港室内设计协会副会长
IFI国际室内建筑/设计师联盟执委

谢健生
室内设计师
香港理工大学设计学院美术及设计学荣誉学士
香港室内设计协会专业会员
担任香港生产力培训学院讲师
2005年成为泛纳国际设计有限公司合伙人

Izakaya Singer
圣家居酒屋

设计单位：PANORAMA泛纳设计事务所
设　计　师：潘鸿彬
参与设计：谢健生、黄卓荣
项目地点：深圳
建筑面积：340m²
竣工时间：2012.07
摄　影　师：吴潇峰

"圣家居酒屋"是深圳市新开辟的饮食和娱乐中心"欢乐海岸"的一间日式餐厅,其主要景观的设计给人"家"的感觉,将传统的日本居酒屋餐厅体验提升到新的潮流水准,将收集起来的各种材料循环使用,提高其用途和观赏价值。

餐区一:

棕色的开放式天花板之下,倾斜的木制屋顶结构在建筑上对传统的日式居酒屋作出了新的诠释。

在铁板烧台面和长木台的上方悬挂着以空米酒瓶和钨丝灯泡装配成的灯饰。不同风格的白色餐椅与木地板成了对比,使整体温暖的色调产生随意但很一致的图案。

餐区二:

由天花吊下的绳帘将餐区分隔成不同的空间。各种不同的座位摆设给食客提供2人,4人,6人及8人的餐位,尚有贵宾房以满足客人所需。在富有特色的墙壁挂满了巨大的黑白色日式烹饪器皿与黄芥末色块拼贴成的装饰,使餐厅独具风格和温馨的品牌形象得以突显及使食客胃口大开。用竹筷编织的吊灯和台灯使整个餐区沉浸在温暖舒适的气氛中。

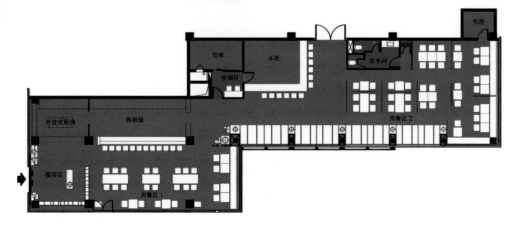

Izakaya Singer is a Japanese restaurant situated at the newly-opened food & entertainment hub "OCT Bay" in Shenzhen, China. A main scene of "Dining Room" was created to give a "home" feeling and move the traditional izakaya's dining experience to a new level of trendiness. Up-cycling of different found objects provide unique casual dining experience for the whole space:

Dining Area 1

Architectural re-interpretation of traditional "Japanese wine house" was given by slanted timber roof structure under brown-painted open ceiling.

Used sake bottles & tungsten light bulbs were re-configured to become pendent lamps above timber bench seatings & Teppanyaki counter respectively.

White dining chairs in various styles set up contrast to the timber floor and provide random yet consistent pattern to the general warm-tone environment.

Dining Area 2

Ceiling suspended rope screens provide flexible compartmentation to the area. Various seating patterns of booth, table of 2 / 4 / 6 / 8 and VIP rooms are introduced to cater for different customer's needs.

Super-sized black & white images of Japanese food cooking utensils collaged with mustard yellow palettes at feature walls create a warm & stylish identity to this new brand and help increased the customer's appetite.

Custom-designed pendent and table lambs made of bamboo chopsticks completed the cozy and warm dining experience.

陈亨寰

来自台湾
大匀国际设计中心协同主持人
台湾中生代设计师
IFI国际室内装饰协会专业会员
中国室内装饰协会专业会员

李巍 Amy Lee

1985年生于中国辽宁省葫岛市
毕业于中国上海东华大学装潢艺术专业

Sanya Qixian Ridge Western Restaurant
三亚七仙岭西餐厅

设计单位：大匀国际设计中心
设 计 师：陈亨寰、李巍
软装设计：上海太舍馆贸易有限公司
项目地点：三亚
建筑面积：1013 m²
竣工时间：2012

明窗延静书，默坐消尘缘

　　日本茶道宗师千利休曾说过："须知道茶之本，不过是烧水点茶"。

　　空间设计本质亦是通过静虑，从平凡的生活中去契悟设计的大道。

　　我们将这个半开放、半私密、半公共的空间溶于天蓝海碧，山清水秀之中。利用庭院、特色走廊等过渡空间，以达到室内外的精美融合。古朴与现代的巧妙混搭，取材天然与自然相融，使人得以在大自然的环境中更惬意体验当下的放空。

　　从来佳茗似佳人，禅茶一味悟自心。1F茶书吧的空间格局被分割成几个细部，更注重于私密空间与公共空间的互动关系。多层次的过厅更给空间增添了仪式感。此外，大面积的户外景观又与室内空间遥相呼应。于此中，若坐若卧，亦是怡然自得。

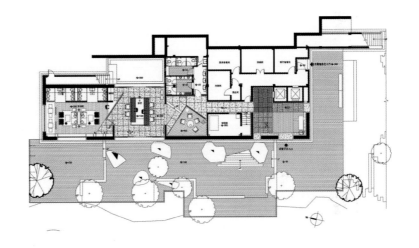

　　其下如是，其上亦然。2F西餐厅空间设计格局更是大面积的运用到落地窗，仿若隐逸空谷中一颗璀璨的钻石。延续1F雕花屏风的局部隔断，光影互动交织成趣，给空间更添私密性。以竹木饰面的吊顶区分整体空间。动线的理性规划，让客户在取餐时更方便。宽大舒适的布艺沙发，更是将整个用餐环境带来了更多的轻松氛围。户外的用餐空间，既是日丽风和或月明星朗的逍遥自在。

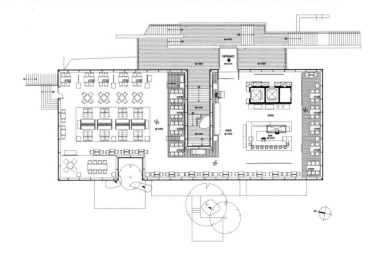

Eading quietly beside the bright window, Sitting silently to eliminate carnal thoughts Sen Rikyu, the master of Japanese tea ceremony, said, "The essence of tea lies in boiling water to make tea".

The essence of space design also lies in understanding the significance of design through meditation and ordinary life.

We blend this semi-open, semi-private and semi-public space with blue sky, azure ocean, green mountain and clear water. We perfectly combine indoor and outdoor space by courtyard and unique corridor. The skillful mix of primitive simplicity and modernism, and the combination of natural materials with the nature enables people to enjoy the more pleasant leisure in the great nature.

The good tea is like a beautiful lady. Therefore, the understanding of Zen and tea is from the heart. Tea and Book House in 1F has been divided into several sections, paying greater attention to interaction of private space with public space. Multi-layer corridor brings a sense of ceremony to the space. In addition, vast outdoor landscape echoes with indoor space from distance, making you comfortable and delightful in sitting or laying down here.

The upstairs is as comfortable as the downstairs. Western Restaurant in 2F has been decorated with large French window, resembling a brilliant diamond hidden in the valley. The partition with engraved screen in 1F has also been applied here to add more privacy to the space with skillful application of lights and shadows. The bambood and wood ceiling separates the whole space. Dynamic and rational partition enables customers to fetch food easily. The large cozy fabric sofas make the dining environment more relaxed. On the other hand, the outdoor dining space is also pleasant with gentle breeze or romantic with bright stars.